Hidden In Plain Sight 10

Andrew Thomas studied physics in the James Clerk Maxwell Building in Edinburgh University, and received his doctorate from Swansea University in 1992.

His *Hidden In Plain Sight* series of books are science bestsellers.

Also by Andrew Thomas:

Hidden In Plain Sight
The simple link between relativity and quantum mechanics

Hidden In Plain Sight 2
The equation of the universe

Hidden In Plain Sight 3
The secret of time

Hidden In Plain Sight 4
The uncertain universe

Hidden In Plain Sight 5
Atom

Hidden In Plain Sight 6
Why three dimensions?

Hidden In Plain Sight 7
The fine-tuned universe

Hidden In Plain Sight 8
How to make an atomic bomb

Hidden In Plain Sight 9
The physics of consciousness

Hidden In Plain Sight 11
The logic of consciousness

Hidden In Plain Sight 12
Consciousness and the steam engine

HIDDEN IN PLAIN SIGHT 10

How to program a quantum computer

ANDREW THOMAS

Hidden In Plain Sight 10

Copyright © 2018 Andrew D.H. Thomas
hiddeninplainsightbook@gmail.com

All rights reserved.

ISBN-13: 978-1726017572
ISBN-10: 1726017575

CONTENTS

PREFACE

The subject of quantum computing is receiving a lot of publicity at the moment, with many stories in the press about how it is going to revolutionise our lives, or break the world's encryption systems. However, most of these stories seem very short on hard facts about how quantum computing actually works. In fact, there seems to be a general lack of easy-to-understand technical information about quantum computing. Whereas there is plenty of complex technical material available.

This book is going to attempt to fill the gap. It will attempt to simplify the subject, clarifying the essential concepts, while still retaining sufficient technical depth.

If you ask in most IT departments about the latest programming languages, or the latest programming techniques, I am sure you would find plenty of programmers with excellent knowledge of the latest developments. However, if you ask about quantum computing, I suspect you would be met with blank stares. But these are the people who – one day, hopefully – are going to be using quantum computers. These are the target customers.

So this book is accessible, aimed at a general audience. But it is also aimed at those programmers and scientific researchers, people with experience of conventional computers who wish to gain knowledge and experience of quantum computing but do not wish to trawl through the extensive and complex literature. This book will explain the programming techniques which will likely be needed to write a quantum computing program to solve their particular problems.

Hopefully, this book can provide another step towards taking quantum computing into the mainstream.

No prior knowledge of quantum mechanics or computing is assumed. This book will take you on the entire journey, from explaining the fundamental postulates of quantum mechanics, then explaining the principles of quantum computing, and finally showing you how you can program a quantum algorithm on a real quantum computer, generously provided by IBM.

You will be working at the frontier of human knowledge, using arguably the most advanced technology which has ever been created. You will be performing complex calculations in a time-window which lasts just microseconds, using single elementary particles which are in two states at the same time, inside a refrigerator which is the coldest place in the entire universe. This is beyond the wildest science fiction.

Quantum computing – and the development of quantum algorithms – is still very much a work in progress. It is a journey into mystery, and no one knows where the destination will be. You can play a part in shaping that future.

Andrew Thomas (hiddeninplainsightbook@gmail.com)
Swansea, UK,
2018

"Get ready to think outside a box you didn't know existed."

- CHARLES BENNETT,
  QUANTUM INFORMATION PIONEER

1

THE THINKING MACHINE

In early 1943, Richard Feynman received a telephone call from Robert Oppenheimer. Feynman was a 25-year-old physicist, a rising star from Princeton University. Oppenheimer was the scientific director of the Manhattan Project, the American project to build the world's first atomic bomb. Oppenheimer persuaded Feynman to join his team and, on March 28th 1943, Richard Feynman boarded a train to Los Alamos, New Mexico, where the centralised bomb development facility was located.

Feynman described his time at Los Alamos in his highly-entertaining autobiography *Surely You're Joking, Mr. Feynman!* He recalled arriving on-site to find the main buildings were still not fully complete, and so the physicists had to lend a hand in the construction work.

Feynman recalls how he attended the introductory lecture given by Robert Serber (which was described in my eighth book) in which the principles behind the atomic bomb were explained. For many of the physicists in attendance, this was the first time they were made aware of the true purpose of Los Alamos.

All of the correspondence in and out of Los Alamos was heavily-censored for fear of spies, and secrets leaking out. This presented a problem for Feynman whose wife, Arlene, was seriously ill with tuberculosis in hospital in Albuquerque. Feynman and Arlene arranged that they would write letters to each other in code. The key to deciphering the code would be written on a separate piece of paper inside the letter.

One day, Feynman received a letter from his wife which included the following key which was written on a separate piece of paper:

LITHARGE
GLYCERINE
HOT DOGS
LAUNDRY

Feynman puzzled over the letter for many hours, unable to decipher it by using the enclosed key. It was only when he arrived at Arlene's hospital a few days later that she told him it was actually a shopping list of things she had wanted him to bring when he visited her.

In another famous anecdote, Feynman is sent on a trip from Los Alamos to inspect the giant uranium enrichment plant at Oak Ridge (also described in my eighth book). When he arrives, his reputation as a genius precedes him, and so the team at Oak Ridge unroll a series of plant blueprint diagrams for Feynman's analysis and approval. However, Feynman has never seen anything like this in his life:

> *I'm completely dazed. Worse, I don't know what the symbols on the blueprint mean! There is some kind of a thing that at first I think is a window. It's a square with a little cross in the middle, all over the damn place. I think it's a window, but no, it can't be a window, because it isn't always at the edge.*

What am I going to do? I get an idea. Maybe it's a valve. I take my finger and I put it down on one of the mysterious little crosses in the middle of one of the blueprints on page three, and I say "What happens if this valve gets stuck?" – figuring they're going to say "That's not a valve, sir, that's a window."

So one looks at the other and says, "Well, if ***that*** *valve gets stuck ..." and he goes up and down on the blueprint, the other guy goes up and down, back and forth, and they both look at each other. They turn around to me and they open their mouths like astonished fish and say "You're absolutely right, sir!"*

So they rolled up the blueprints and away they went and we walked out. And Mr. Zumwalt, who had been following me all the way through, said: "You're a genius."

At Los Alamos, Feynman was assigned to the team responsible for analysing the behaviour of the bomb during implosion – the period during which a sphere of plutonium about the size of an orange is crushed to about the size of a lime. The task of the team was to calculate how much explosive energy would be released during the implosion. However, they quickly realised that this required more calculating power than they had available.

The physicists at Los Alamos were also all too aware that their work was a race against time. They had to beat Hitler to the bomb – the alternative was too awful to contemplate. Because of those extreme time pressures, Feynman realised he had to find a way of reducing the time of their calculations. Some form of automation was going to be required. However, those were the days before the digital computer was invented (during the same period, the digital computer was being invented in Britain in order to crack the German Enigma code).

However, IBM had produced a series of mechanical calculating machines which could perform a limited set of operations on punched cards. Each machine was only capable of performing a single type of operation, such as multiplying two numbers on a punched card together. As an example of one of these type of machines, the following photograph shows the IBM 601 Multiplier, first introduced in 1931, which is known to have been the type used by Feynman in Los Alamos. This was the first IBM calculator which could multiply. You can see the stack of iconic 80-column IBM punched cards in the top right of the photograph:

According to the IBM website, the punched card running on machines such as this:

> *was the dominant form of data processing from 1890 until commercial electronic computers arrived in the 1950s. That's more than a half-century of transforming business in virtually every industry in the world. Machines such as this made IBM into one of the*

> *few major corporate success stories of the Great Depression, and launched the company on its path to becoming a computing giant.*[1]

However, with each machine only capable of performing a very limited number of different operations – or maybe only one type of operation – any complex operation would require a number of these machines to be placed in a line. The cards output from one machine would then be fed into the next machine in the line – very much like an industrial assembly line. Feynman described this approach:

> *If we got enough of these machines in a room, we could take the cards and put them through a cycle. Everybody who does numerical calculations now knows exactly what I'm talking about, but this was kind of a new thing then – mass production with machines.*

Back in Los Alamos, with each calculating machine performing just one particular operation, this represented a single "assembly line" with each stage being performed in sequence. In modern computing terms, this would represent what we now call *serial processing.*

The following diagram shows Feynman's serial processing arrangement at Los Alamos, with punched cards entering at the top, processed in a series of stages (mechanical calculators), and exiting with the result of the calculation at the bottom of the diagram:

---

[1] **http://tinyurl.com/ibmpunchedcard**

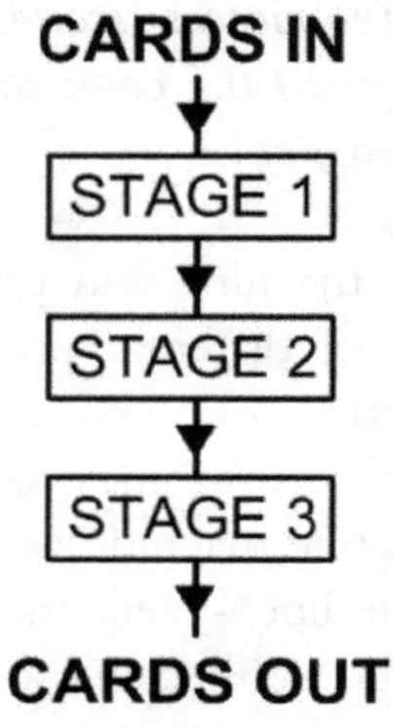

SERIAL PROCESSING

In a mass production system, it is possible to increase output just by adding more assembly lines: the more lines you add, the more items you can manufacture in a day. This represents a move from a **serial** assembly process to a process in which there are many **parallel** lines. As a consequence, the average manufacturing time for each item can be greatly reduced. Feynman applied a similar principle to his serial processing arrangement at Los Alamos.

The way Feynman achieved this was ingenious. He still used a single processing line of machines, but he used cards which were different colours. In that way, he could have many cards being processed on the single line at the same time, with the colours of the cards allowing him to group the cards which were at were at different stages of the process.

Here is how Feynman described this approach:

> *The problems consisted of a bunch of cards that had to go through a cycle. First add, then multiply – and so it went through the cycle of machines in this room, slowly, as it went around and around. So we figured a way to put a different coloured set of cards through the cycle too, but out of phase. We'd do two or three problems at a time.*

Effectively, Feynman had created multiple virtual processing lines which ran in parallel, as shown on the following diagram:

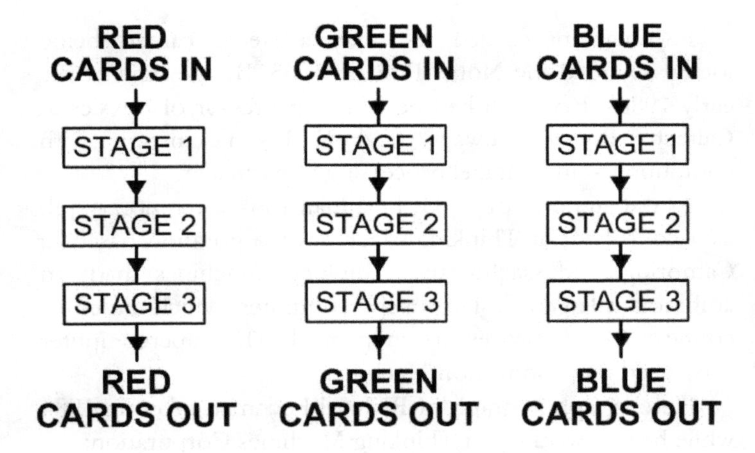

PARALLEL PROCESSING

As shown in the previous diagram, this approach represents *parallel processing*. Parallel processing is a computing technique which allows large amounts of data to be processed in a much shorter overall length of time. And it was Feynman's experiments in parallel processing at Los Alamos which planted the seed for the quantum computer in his mind.

In this book we will be examining how a quantum computer uses parallel processing techniques to achieve huge increases in processing speed. And – in honour of Feynman's experiments at Los Alamos – I will be retaining the colour idea ("red", "blue", etc.) in the examples in this book to describe the items being processed in parallel.

Feynman and the Thinking Machine

Feynman proceeded to enjoy a stellar career, being jointly-awarded the Nobel Prize in 1965. By the time of the early 1980s, Feynman had become a professor of physics at Caltech. But he always retained his fascination with computing – and parallel processing in particular.

In the summer of 1983, Feynman took a temporary job as a consultant at Thinking Machines Corporation based in Cambridge, Massachusetts. Thinking Machines had an ambitious plan to design a supercomputer, which aimed to connect 64,000 processors in parallel. The supercomputer was called the Connection Machine.

Here is a photograph of Richard Feynman taken in 1984 while he was working at Thinking Machines Corporation:

Connecting 64,000 processors in parallel might sound like an impossible task, but we will see later in this book that a similar amount of parallelism could be achieved by just 16 quantum bits (because 2^{16}=65,536).

The company was being set-up in an old mansion when Richard Feynman arrived just a day after the company was incorporated. The situation was still in a state of confusion when Feynman arrived, saluted, and said: "Richard Feynman reporting for duty. OK, boss, what's my assignment?"

After a quick private discussion ("I don't know what he can do – you hired him!"), they suggested that Feynman could act as a technical advisor, to which Feynman responded: "That sounds like a bunch of baloney. Give me something real to do."

At which point they sent him out to buy office supplies.

Each processor in the Connection Machine was fairly simple, but the network was complex: each processor was connected to twenty other processors. Feynman was given the job of designing the router, which controlled the traffic flow of data between the processors. The other team members were just glad that they had found something to keep Feynman occupied – or so they thought. Because, in his spare time, Feynman couldn't help wiring the computer room, setting-up the machine shop, and installing the telephones. He also organised the company into teams, relying on his Los Alamos experience: "We've got to get these guys organized. Let me tell you how we did it at Los Alamos."

Feynman revealed the potential of the Connection Machine for simulating aspects of the physical world in which many particles interact at the same time. The example chosen by Feynman was the interactions between the particles which constitute the protons and neutrons in the atomic nucleus. Protons and neutrons are made from elementary particles called *quarks*, and quarks are held together by the *strong force*. The particles which transmit the strong force between the quarks are called *gluons*.

Gluons not only interact with quarks – they also interact with other gluons. Another complicating factor is introduced because the strong force is highly nonlinear. If you read my previous book, you will recognise that when many elements interact in a nonlinear fashion then this represents a system with high complexity. Another example of a nonlinear system is the weather, caused by the nonlinear behaviour of the turbulent atmosphere. Nonlinear systems are notoriously difficult to predict, and there is usually no other option than to split the problem into small pieces and then to use extremely powerful computers to plot the behaviour of those small pieces over time.

As an example of that technique, in order to predict the weather, meteorologists analyse the turbulent and chaotic atmosphere by splitting it into many small pieces maybe a few kilometres apart. A similar principle is used to predict the behaviour of quarks and gluons in the atomic nucleus. The space inside a proton or neutron is split it into a three-dimensional lattice of points, each point in the lattice representing a different point in space. The lattice can then be mapped onto a parallel processing supercomputer, with each processor in the supercomputer representing a point in space: the finer the lattice, the more accurate is the simulation.

This method is called *lattice QCD* (the theory of the strong force interactions is *quantum chromodynamics*, or QCD).

Feynman used this technique of lattice QCD to map each point in the lattice to a processor of the Connection Machine. He was delighted to discover that the Connection Machine could outperform the conventional computer that Caltech was using for QCD calculations. Once again – as in Los Alamos – parallel processing had proven its ability to tackle a huge volume of data in a reasonable time.

But this got Feynman thinking.

It was clear that a massively parallel computer could outperform a conventional computer in creating simulations of physical systems. This was because the supercomputer was a more accurate representation of how Nature actually works: many small units (particles) interacting in parallel. In that case, wouldn't it be possible to build an even faster and more accurate computer by making it a **perfect** match of Nature, in other words, build a computer which used individual particles as its data elements.

The idea for the quantum computer was forming in Feynman's mind.

The Feynman Lectures on Computation

From 1983 to 1986, Feynman gave a course on computation at Caltech. At the end of the course, Feynman asked one of his colleagues at Caltech, Tony Hey, to adapt his lecture notes into a book. That book is now called *The Feynman Lectures on Computation.* We are lucky that we now have that book generally available so we can follow Feynman's remarkable insights.

The course represented a very different approach to how we now think of "computers" or "computer science". If you browse the computing section of a bookshop, you might find many books on high-level languages such as Visual Basic or C++. Or in the magazine department you might find WIRED magazine, giving you the latest "tech" news about Snapchat or Facebook. For many people, this is what "computing" means nowadays – it's nothing more than fashion, basically.

Feynman, however, considered a very different type of "computing" in his course. Feynman considered the lowest-level of computing, the actual movement of physical material in order to perform a calculation. Feynman considered fundamental physical concepts such as energy, information, and thermodynamics. As Tony Hey says in the foreword of the lecture notes:

> *As advertised, Feynman's lecture course set out to explore the limitations and potentialities of computers. Although the lectures were given some ten years ago, much of the material is relatively timeless and represents a Feynmanesque overview of some standard topics in computer science. Taken as a whole, however, the course is unusual and genuinely interdisciplinary. Besides giving the 'Feynman treatment' to subjects such as computability, Turing machines (or, as Feynman says, 'Mr. Turing's machines'), Shannon's theorem and information theory, Feynman also discusses reversible computation, thermodynamics, and quantum computation. Such a wide-ranging discussion of the fundamental basis of computers is undoubtedly unique and a 'sideways', Feynman-type view of the whole of the subject.*

(In that quote, Tony Hey mentions "reversible computing" which we will be considering in detail later in this book.)

Because of his interest and aptitude for fundamental science, it appears that Feynman was one of the first people to realise the potential of quantum computing. He was certainly the first person to introduce the principles to a wider audience in his lecture course.

Feynman's proposal for a quantum computer is unveiled in Chapter Six of his book. Feynman's motivation in considering quantum computing at this stage seems to be to treat it merely as an intellectual exercise. Specifically, Feynman seems to be fascinated by how small it might be possible to build a working computer: could just a few particles be made to represent a computer? In that way, Feynman is following a train of thought which started back in 1959 in one of his famous lectures called *There's Plenty of Room at the Bottom* which considered the possibility of creating microscopic machines made of just a few atoms. That lecture was to inspire the field of *nanotechnology*. Feynman's motivation in suggesting quantum computing seems to be merely an intellectual challenge, asking interesting questions, and having a bit of fun in the process. It is classic Feynman.

And, at the end of Chapter Six, Feynman comes to the following amazing conclusion:

> *It seems that the laws of physics present no barrier to reducing the size of computers until bits are the size of atoms, and quantum behavior holds dominant sway.*

Feynman had come to his conclusion: quantum computers **were** possible.

The classical computer

However, in his book, Feynman starts by introducing the principles which lie behind the conventional classical computer. "Classical" physics means the state of physics as it existed before 1900, which means it pre-dates the discovery of quantum mechanics and relativity. It is the physics of Newton – not the physics of Einstein and Bohr. Classical physics is also the physics of the great Scottish 19th century physicist James Clerk Maxwell who viewed light as a wave rather than a stream of particles. The computers we use today are all classical computers, based on classical physics.

The template for the classical computer is the *Turing machine*. The Turing machine was invented by Alan Turing in 1936, and it is the simplest possible general computing device. The Turing machine operates by shuffling a long strip of paper back and forth. The tape has a series of zeroes and ones written on it, representing input data and output data. The Turing machine has a probe pointed at the paper which can read, write, or erase the zeroes and ones on the strip:

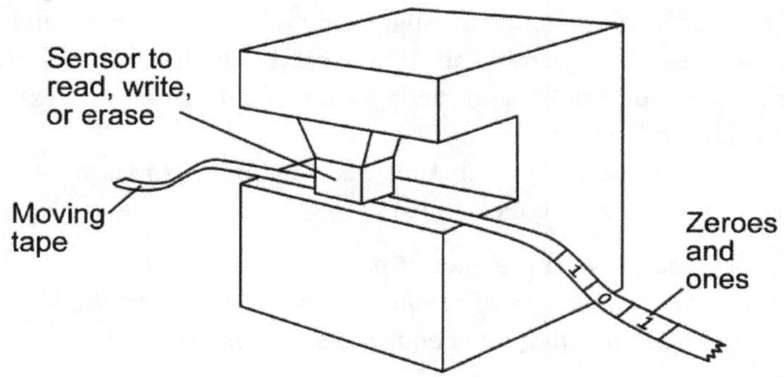

The machine has an in-built table of rules – the equivalent of a computer program – which tell it what to do depending on the current symbol on the tape and the current state of the machine. The initial set of symbols on the tape would represent the input to the machine, and the final set of symbols – after processing – would represent the output.

So a Turing machine is a computer. But the most remarkable fact about the Turing machine is that **it can solve any problem which can be solved by any other computer.** This is because any computer – no matter how apparently sophisticated it might appear – is, at heart, merely shuffling individual zeroes and ones around according to a list of instructions (a program) just like the Turing machine. Admittedly, other computers might work a lot faster, or might have prettier screens on which to output their data, but the basic problem-solving capability of any computer is just the same as a Turing machine.

Put simply, a Turing machine can compute anything that can be computed.

It is for this reason that the Turing machine forms the model for every modern computer – including the PC on your desk, or your iPhone or iPad. There are some implementation differences: a modern computer writes data to-and-from electronic memory, whereas a Turing machine writes data to-and-from a tape. However, in both cases, only one item of data is read from memory and processed at a time. Wouldn't it be faster if it could work in parallel, processing many data items simultaneously, like Feynman's machine at Los Alamos?

Yes, indeed it would. And that is the promise of quantum computing.

A classical computer is based on classical physics. As you can see from the previous diagram, it is basically a clunky thing. It is based on pre-20th century physics, the physics of Newton and James Clerk Maxwell. A Turing machine – the basis of the modern computer – could have been built in the year 1899 from the technology in existence at that time. To emphasise this point, see the amazing video on YouTube of the Turing machine made of Lego:

http://tinyurl.com/plasticmachine

or the Turing machine made of wood:

http://tinyurl.com/woodmachine

These contraptions might seem crazy, but they can compute anything your iPhone can compute (at least, in theory).

So the modern digital computer is actually rather old-fashioned. It is, at its heart, a 19th century device. It needs updating for the 21st century. What it really needs is to be based on the extraordinary 20th century phenomenon of quantum mechanics.

Put simply, it needs to become a computer which you can't make from Lego.

It needs to become a quantum computer.

2

THE QUANTUM STATE (OF MIND)

Many books aimed at a general audience have been published attempting to explain the subject of quantum mechanics. The majority of those explanations tend to follow a fairly linear historical series of events, starting in the year 1900 and ending a few decades later. It is a neat story, featuring some of the greatest physicists in history, revealing their heated arguments and strong personalities as they slowly edged their way towards understanding.

Yes, it is a great story. But it is probably a bad way to learn about quantum mechanics.

In his recent book *Beyond Weird*, Philip Ball agrees: "The temptation to tell quantum mechanics as a historical saga is overwhelming". But this is a temptation which should be resisted.

The problem is that for the first two decades of the 20th century, the field known as "quantum theory" – or now known as the "old quantum theory" – was a mess. Nothing made sense. A series of puzzling experimental results had shown that the classical physics of the 19th century could not be entirely accurate. However, the attempts to bring the old physics into line with the new experimental results just resulted in a series of hodge-podge theories which were soon to be discarded. As Philip Ball says: "They cobbled old concepts and methods together". The Wikipedia page on the "old quantum theory" describes the situation as:

> *The old quantum theory is a collection of results from the years 1900–1925 which predate modern quantum mechanics. The theory was never complete or self-consistent, but was rather a set of heuristic corrections to classical mechanics.*

As Philip Ball continues to say: "There is no reason to believe that the most important aspects of the theory are those that were discovered first, and plenty of reason to think that they are not." Even the name "quantum theory" which arose from this period is a bad choice of name. As Philip Ball says: "If we were naming it today, we'd call it something else".

No, if we want to learn about quantum mechanics, then we need to skip forward in time to the mid-1920s. That is because, in 1925, everything changed – and a new light of understanding and clarity was cast over the subject.

In 1925, the "old quantum theory" was superseded by the theory of quantum mechanics, which was separately developed by Werner Heisenberg and Erwin Schrödinger. This was a theory which brought together all the puzzling, arbitrary rules of the old quantum theory into a single, logical theoretical framework. In doing so, quantum mechanics replaced the old classical mechanics.

The old classical model not only had to be replaced by a new theory of mechanics – it also had to be replaced by a new theory of reality! Quantum mechanics represented more than a theory about the behaviour of atoms: it represented a new, more sophisticated way of thinking about reality and how the world must work. Though you might sometimes read of quantum mechanics being described as "weird", I hope to show you that nothing could be further from the truth: quantum mechanics is entirely sensible and logical.

But, more than that, I want to show you that quantum mechanics is more than a theory – it is a new way of thinking about reality, a better, more logical way of thinking about reality.

I want to show you that quantum mechanics is a state of mind.

The new worldview of quantum mechanics

What is reality? From a scientific viewpoint, we can define reality as being the values of the properties of objects, such as particles. For example, if we can fully describe the properties of all the particles which comprise a system, then we have a full definition of the reality of that system.

Over the next few chapters we will be seeing that quantum mechanics considers reality (for example, the property values of a particle) as being divided into two periods of time:

1) The reality of the particle **BEFORE** it is observed.

This is not as straightforward as it might sound, raising deep questions about "reality-without-observation" which resemble the old philosophical thought experiment: "If a tree falls in the forest and no one is around to hear it, does it make a sound?"

2) The reality of the particle **AFTER** it is observed.

We shall be seeing that the determination of the particle values when it is observed (measured) involves fundamental uncertainty and randomness which is unavoidably introduced by quantum mechanics.

While this new version of reality revealed by quantum mechanics might seem rather strange at first, I hope to show you that it makes perfect sense, and – when you think about it – you will realise it represents a rather more sophisticated model of how reality must work.

OK, let's start with a simple model of how the classical world was imagined:

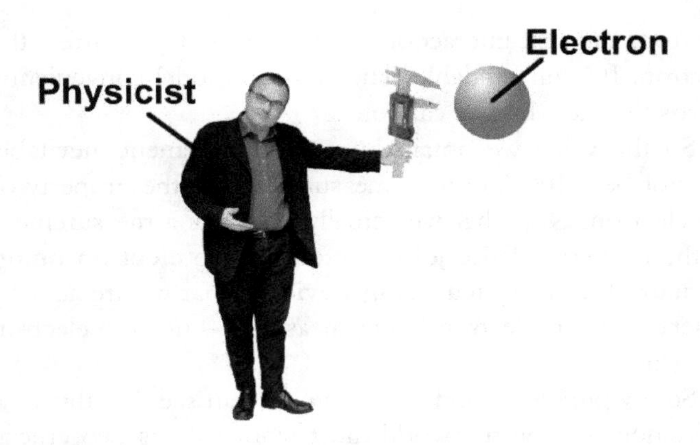

The previous diagram shows a physicist holding a highly-sophisticated measuring device. He is going to use his device to measure some property of the electron (not shown to scale).

According to the classical view, this is a fairly straightforward operation. The measuring device and the electron are two clearly defined, separate objects. The electron has a definite reality, in other words, not only does it definitely exist but it possesses definite values for its position, speed, etc. The physicist then uses his measuring device to "extract" those values from the electron, and writes them down. Problem solved.

Only, when you think about it, you realise this can't be a correct picture of what actually happens.

For example, let us imagine the physicist wants to measure the electron's position. As an electron moves, it does not continuously (and conveniently) radiate data about its position. No, if we want to determine the position of the electron, we have to observe it, or interrogate it. Basically, we have to ask it a question. That observation process might involve shining a light on it (like a prisoner undergoing an interrogation), or forcing the electron to come to a rest by hitting a screen and then noting its position. But whatever method we use, we are going to have to interact with the electron, and that interaction is inevitably going to affect the electron. It is unavoidable. The classical model conveniently avoids that fact, but we cannot.

So the value we obtain by our measurement inevitably can not be a 100% perfect measurement of the property of the electron. No, what we actually obtain is a measurement of the property of **the joint system of the electron being measured by the measuring device.** What we are actually observing is an electron being measured – not the electron on its own.

So, by pure logic and reasoning, we can see that this new realisation of how the world must work tells us two crucial facts about reality, which have to be incorporated into the theory of quantum mechanics:

- A measurement inevitably affects the object under observation. We are therefore not measuring the true reality of the object on its own, i.e., the object before it is measured.
- Objects are never completely isolated. The measurement we eventually obtain is a measurement of the joint system: "an electron being measured by a measuring device".

What is more, these principles do not just apply to quantum mechanics. These are general principles which apply to all aspects of our lives. No object is ever completely isolated, and all actions affect the underlying reality.

A rather unusual example of our actions affecting the underlying reality was provided by the billionaire investor George Soros. Soros named his investment fund the Quantum Fund when he realised that his own purchasing actions were inevitably affecting the prices of the stocks and shares he was buying.

The quantum philosophy is a logical philosophy which affects many aspects of our lives.

The quantum state

Let us return to consider the first of the two principles presented in the previous section. It was stated that:

> *A measurement inevitably affects the object under observation. We are therefore not measuring the true reality of the object on its own, i.e., the object before it is measured.*

So when we make a measurement of some property of a particle, all we can ever hope to achieve is to measure the state of the particle after it has been measured, after the measurement process has fundamentally altered the particle in some way. So what can we say about the reality of the particle **before** it is measured? Surely that is what we really want to capture. Surely that represents the true reality of the particle.

Well, yes, and even though we can never capture that reality of the particle through measurement, we **can** reveal it through mathematics. Mathematically, the state of a system before it is measured is called the *quantum state*, and is conventionally denoted by the Greek letter Ψ (written "psi" and pronounced "sigh").

Sometimes this quantum state is referred-to as the *wavefunction*, with the Ψ symbol being especially associated with the wavefunction. It is called a wavefunction because the physical nature of the quantum state will often resemble a wave (such as the path of an electron around an atomic nucleus, as we shall see later in this book). However, the terms "quantum state" and "wavefunction" both mean essentially the same thing, though I will be preferring the term "quantum state" in this book.

This principle – that Ψ represents the physical reality of the system before it is measured – is expressed several times by the physicist Roger Penrose in his popular science books such as *The Emperor's New Mind*: "I am taking the view that the physical reality of the particle's location is, indeed, its quantum state Ψ", and again in *The Road to Reality*: "If we are to believe that any one thing in the quantum formalism is actually real, for a quantum system, then I think that it has to be the wavefunction (quantum state) that describes quantum reality." I would agree with Roger Penrose that we should consider the quantum state as representing the reality of a quantum system before observation, and I would encourage you to do the same. This is especially true when, later in this book, you will find that your time spent programming a quantum computer involves rotating a quantum state vector – which makes it feel very "real" indeed!

To understand the nature and behaviour of Ψ, let us draw a graphical representation of what happens when we measure some property of a particle. To make things simple, I will use the purely imaginary notion that a particle has some form of visual "colour" of some kind, and when we inspect that particle we will find it to be either "red" or "blue" (in honour of Feynman's coloured cards at Los Alamos).

The following diagram shows the graph. You will see that the vertical axis represents the "amount of red" in the particle, and the horizontal axis represents the "amount of blue" in the particle:

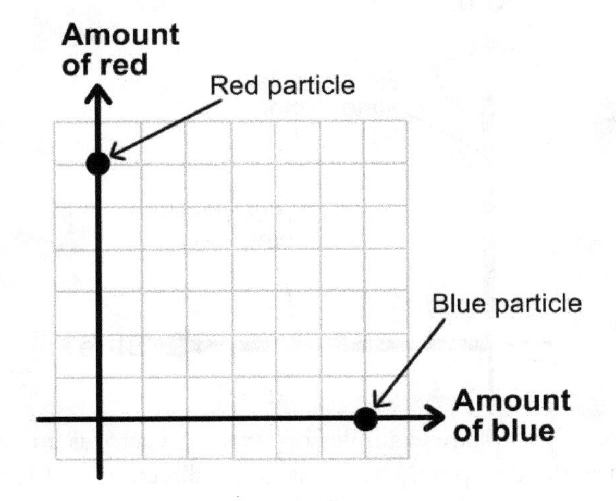

As you can see on the previous diagram, the red particle will have "a lot of red" and "no blue at all", so we plot it high on the "amount of red" axis, but at the extreme left (the zero position) on the "amount of blue" axis. Similarly, the blue particle will have "a lot of blue" and "no red at all", so we plot it on the far right of the "amount of blue" axis, but at the zero position (at the bottom) of the "amount of red" axis.

We can then see that what we have here are two dots plotted which are both the same distance from the origin of the graph (the origin is the zero position of the graph, where the two thick axes cross at the bottom left). So the best way to plot these points is as arrows which have equal lengths, coming out of the origin of the graph. The points representing the particle states (colours) would then be at the end of the arrows:

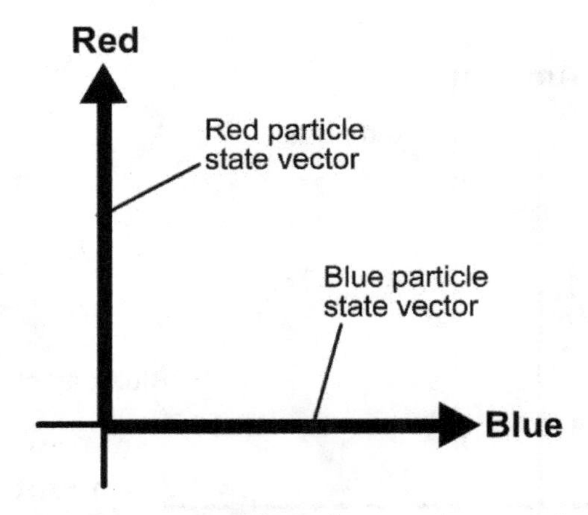

These arrows are examples of *vectors*. A vector is an arrow which is defined purely in terms of its direction and length. In other words, if two arrows have the same direction and length then they represent the same vector – **even if they are positioned at different points in space.** Please make sure you understand this vitally important principle about a vector: all that matters is the direction and length of a vector – it does not matter where it is.

A vector is usually written as a column matrix, which simply represents the coordinates of the endpoint of the arrow. For example, the red vector arrow in the previous diagram has an endpoint which is some distance up the vertical axis, but has a zero horizontal coordinate, while the blue vector arrow has an endpoint with a zero vertical value and a non-zero horizontal value. Assuming each vector has a length of one unit (which is usual), that means the two vectors could then be written as the following column vectors:

$$\text{red} = \begin{bmatrix} 0 \\ 1 \end{bmatrix} \quad \text{blue} = \begin{bmatrix} 1 \\ 0 \end{bmatrix}$$

The particular vector which describes the current quantum state of a system is called the *quantum state vector* or simply the *state vector*. We will be encountering these arrows – these “state vectors” – very often in our discussion of quantum computing. We will see that they are absolutely central to understanding how quantum computing works. We will see that programming a quantum computer is basically the act of manipulating a state vector.

But these mathematical vectors are not arrows in the real world – they are not pointing from point A to point B in physical space. No, they only exist as mathematical constructions drawn on a graph. It might help you to visualize them as physical arrows in physical space, but do not make the mistake of thinking that that is their true nature.

So these vectors live in a mathematical world. They are mathematical constructions drawn on a graph. And that particular type of graph is called a *vector space*.

And, now we have moved inside a mathematical world, we have to realise that these vectors behave according to mathematical rules – not physical rules. And one of the rules of a vector space is that we can combine, or add, any two vectors together to make a third vector.

So, how do we add two vectors? Well, remember that a vector is defined as a length in a certain direction. So if we want to add two vectors, we might think of walking along the first vector, continuing in that direction to the end of that first vector, and then from that point continuing to walk along the second vector, as shown in the following diagram:

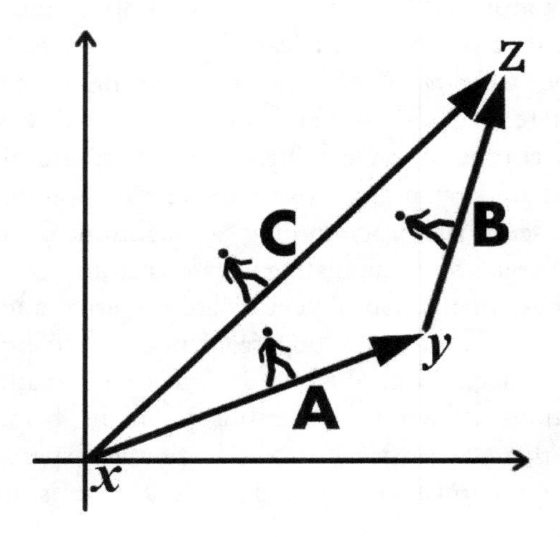

The previous diagram shows how vector **A** and vector **B** can be added to produce vector **C**. It is equivalent to walking along vector **A** from point *x* to point *y*, and then adding vector **B** by walking a further distance from point *y* to point *z*. As you can see from the diagram, the end result will be as if you had simply walked from point *x* to point *z* directly, so we say that vector **C** (the direct route from point *x* to point *z*) is the result of adding vector **A** and vector **B**.

The following diagram shows a slightly different geometrical method of obtaining the same result. You can again see exactly the same vectors **A** and **B** as in the previous diagram. They are the same vectors as in the previous diagram as their lengths and directions are the same as in the previous diagram – even though vector **B** is not positioned in the same place as in the previous diagram. Remember: as explained earlier, if a vector has the same length and direction as another vector, then they are the same vector – no matter if they are positioned at different points in space.

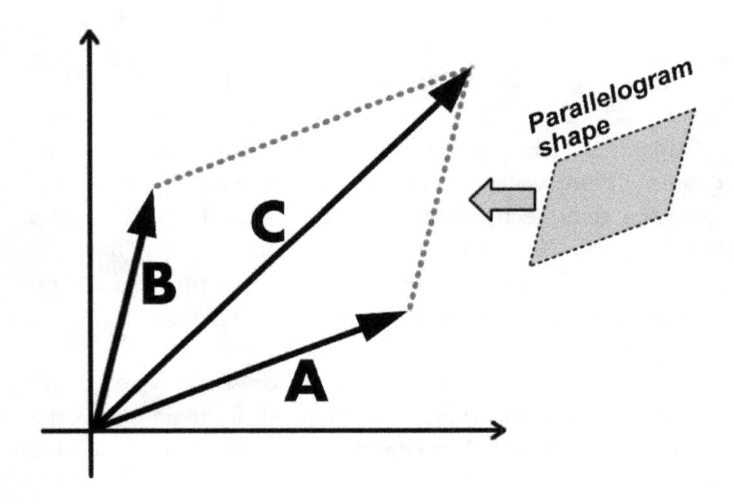

You can see in the previous diagram that this second vector addition method involves the drawing of a parallelogram, with vectors **A** and **B** forming two sides of the parallelogram. The result of the addition – vector **C** in this case – is then the diagonal of the parallelogram. You will also see that vector **C** – the result of the addition – is exactly the same as in the previous diagram, so both methods give exactly the same result.

It is this second method – using the parallelogram – which is the most common geometric method for visualising the addition of two vectors. In fact, the method of vector addition is called the *parallelogram law*.

The superposition state

Reality at the quantum level is therefore defined by the quantum state vector. And, as has just been explained, the quantum state vector lives in a mathematical world called a vector space. One of the rules of a vector space is that it is possible to add any two vectors together to produce a third vector – and we have just seen how this is possible.

But this leads to an intriguing conclusion.

If, according to the rules of a vector space, we can add any two vectors together, then let us apply that rule to the vector space presented earlier which described a red and blue particle:

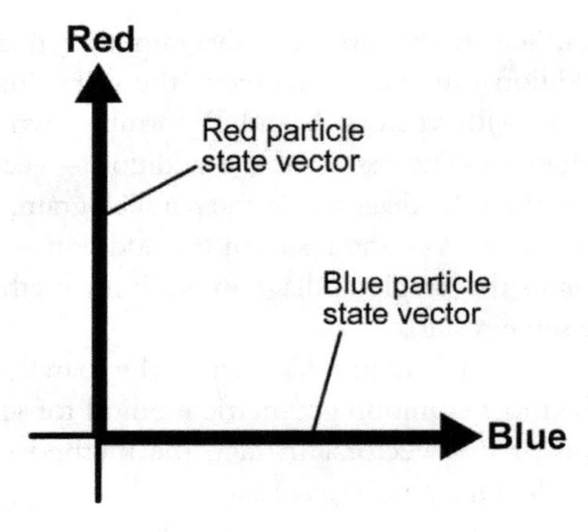

By the rules of the vector space, we can apply the parallelogram rule to add those two vectors, thereby producing a new, third vector:

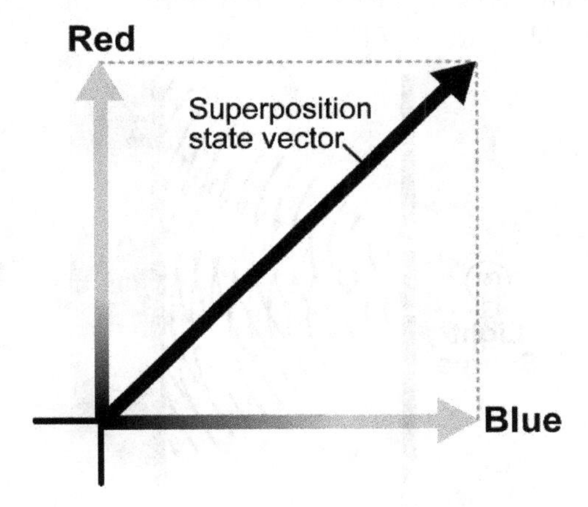

As you can see in the previous diagram, the parallelogram rule has been applied, thereby adding the two state vectors. As a result, a third vector has been generated which represents the addition of the red and blue state vectors. It is as though the particle has a mix of properties: as though it is both red and blue at the same time.

Initially, this might not seem like a particularly remarkable result. After all, colours can be mixed together (when mixing paint, for example). However, we have to remember that other properties of a particle – for example, its position or momentum – are also described by state vectors in vector space. In that case, how on Earth could a particle have two values of position which are "mixed together"? Can a particle really be in two places at once? This appears to be what the mathematics of vector spaces is telling us.

Well, yes, amazing though it may appear, sometimes a particle can behave as though it is in two places at once. The most famous experiment revealing this behaviour is the double-slit experiment in which a beam of light is directed towards two narrow slits, with a screen positioned on the other side of the slits. The light beam passes through both of the slits, generating an interference pattern on the screen due to the wavelike nature of light:

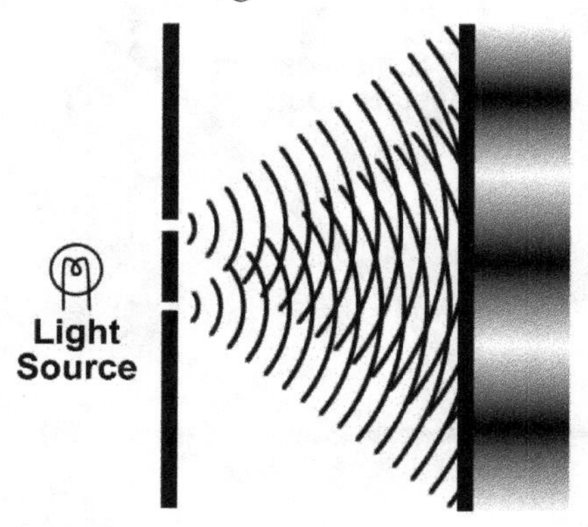

The interference pattern is generated by light passing through both of the slits, with the light from one slit interfering with the light passing through the other slit. This behaviour is to be expected according to classical physics (James Clerk Maxwell's classical theory of electromagnetism describes light as a wave).

However, now let us consider the situation when the light intensity is reduced so that only a single photon (a photon is a particle of light) is in transit at a time. Over time, a pattern of marks will develop on the screen as the individual photons strike the screen and accumulate. Amazingly, though, over time, the interference pattern is still produced! How can this be, when there is only one photon in the system at a time? It is as if the single photon passes through both slits and interferes with itself! As amazing as this may appear, we have just seen that this is behaviour which is predicted by quantum mechanics and the mathematics of vector spaces (the addition of state vectors).

When a particle has a mix of properties in this way, it is said to be in a *superposition state*. We shall be seeing that the ability of particles to enter a superposition state plays a central role in quantum computing.

Quantum mechanics is not "strange"

Throughout this chapter I have been stressing that we can gain an understanding of quantum mechanical behaviour through mathematics and logical reasoning. Earlier in this chapter it was explained via logical reasoning how a measurement will inevitably affect the object being measured – one of the central principles of quantum mechanics. Then, in the previous section, it was explained how the superposition state – which many would consider to be "strange" – is, in fact, predicted by the rules of mathematics. **In fact, it would be stranger if the superposition state did not exist!** (How can you have a world of mathematics in which you can't add things up?)

And this makes an important point: it is not the quantum mechanical world which is strange – if anything, it is our human-scale ("macroscopic") world which is strange! If there is a puzzle at all in quantum mechanics, it is the puzzle of why we do not see quantum superpositions in everyday objects – why we do not see Schrödinger's famous cat both alive and dead at the same time.

This is a viewpoint which is nicely expressed in the IBM Quantum Experience full user guide (we will be programming the IBM quantum computer later):

> *By making quantum concepts more widely understood – even on a general level – we can more deeply explore all the possibilities quantum computing offers, and more rapidly bring its exciting power to* ***a world whose perspective is limited by classical physics.***

It is an excellent piece of writing. I have placed the phrase "a world whose perspective is limited by classical physics" in bold, and I would especially like to emphasize the word "limited". The macroscopic world – the classical world

– in which we live appears to arise due to limitations which prevent us from seeing objects in superposition states. It would appear there is some strange and unusual mechanism at work which filters-out those superposition states, leaving us with a rather false view of how Nature operates.

In fact, in Chapter Six we will be seeing that this "strange and unusual mechanism" is achieved through a fairly well-understood physical process called decoherence. We will also see that the randomness which plays such a fundamental role in quantum mechanics gets "averaged-out" by decoherence, leaving the human world based on deterministic cause-and-effect which we all know so well. We will also be seeing that it is only by preventing decoherence that we can avoid those limitations of classical physics, thereby unleashing the full power of quantum mechanics in our quantum computers.

In this respect, it is the human world which is peculiar, a rather fake and illusory version of the true quantum reality. As physicists, we must be very careful not to let our human intuition bias us into deciding what is "weird" and what is "sensible". And yet there are still some writers who insist that the quantum world is somehow "weird" – with presumably the human world supposed to represent some paragon of sense and predictability.

The human world is not weird? I mean, seriously? When was the last time they took a ride on the subway …

Dirac notation

This is the last section in this chapter describing the quantum state. Before this chapter ends, another important piece of notation needs to be introduced. This notation is commonly used whenever the quantum state is considered, and it is used throughout most technical discussions on quantum computing.

Paul Dirac was unquestionably the greatest British physicist of the 20th century. Dirac was one of the pioneers of quantum mechanics, being awarded the Nobel Prize in 1933 together with Erwin Schrödinger – another of the great quantum pioneers (Werner Heisenberg had received his Nobel Prize the previous year). Dirac was a socially-awkward, taciturn character. You can read about his story and work – especially the remarkable *Dirac equation* – in my fifth book which is about particle physics.

One of Dirac's most enduring contributions was by creating the standard form of notation used for describing quantum states. According to the Dirac notation, a quantum state is denoted by simply placing the name of the quantum state inside angled brackets in the following form:

$$|\text{red}\rangle$$

or some other common examples:

$$|0\rangle \text{ or } |1\rangle \text{ or } |\text{OFF}\rangle \text{ or } |\text{ALIVE}\rangle$$

When a state vector is described like this it is called a "ket" (as part of Dirac's "bra-ket" notation).

You will encounter this notation throughout quantum mechanics and quantum computing theory – and it is used in the remainder of this book.

3

THE SCHRÖDINGER EQUATION

In the previous chapter, the quantum state vector was introduced. We will see later in this book that programming a quantum computer involves rotating the state vector in various ways, to point in different directions. Each movement of the quantum state vector represents a step towards the completion of a calculation on a quantum computer.

In this chapter, we will be seeing that this rotation of the state vector is actually a well-defined process, with no uncertainty about how the state vector moves. And we will be seeing that the movement of the state vector is described by one of the most famous equations in physics.

The life and loves of Erwin Schrödinger

Erwin Schrödinger was an Austrian physicist who was one of the most important pioneers of quantum mechanics.

Schrödinger was born in Vienna in 1887, and became an exceptionally gifted pupil at school. According to one of his fellow pupils: "Schrödinger had a gift for understanding that allowed him immediately and directly to comprehend all the material." Schrödinger received his doctorate in physics from the University of Vienna in 1910, and became a professor of theoretical physics at the University of Zurich in 1921.

Outwardly, then, Schrödinger lived a highly-respectable and conventional life. However, Schrödinger and his wife Annemarie had an open marriage, with both having a string of affairs. The situation was complicated, to say the least. Annemarie had an affair with the physicist Hermann Weyl, who happened to be a good friend of Erwin Schrödinger. Meanwhile, Erwin Schrödinger was working with his assistant, Arthur March, because he was sleeping with Arthur's wife, Hilde, and eventually fathered a child with her.

By this stage, Annemarie, seems to have become aggrieved by Schrödinger's constant infidelities, but she remained attracted to his intellect and decided to stay in the marriage, saying: "You know it would be easier to live with a canary than a racehorse, but I prefer the racehorse."

Schrödinger persuaded Annemarie to allow Hilde to live in their house, forming an uncomfortable long-term *ménage à trios.* However, living with two women was not enough for Erwin Schrödinger, who had two other girlfriends and had children with each of them.

Here is a photograph of Erwin Schrödinger. I must say, I think he's looking rather worn-out …

The Schrödinger equation

We will now consider Schrödinger's breakthrough work of 1925, the development of an equation which effectively created the field of quantum mechanics.

Let us imagine we want to observe the value of some property of a quantum system – for example, measuring the momentum or position of a particle (in quantum mechanics, a measured property is called an *observable* of the system). The first thing to be realised is that if we want to measure a certain property then we need to apply a certain measurement operation, for example, we might measure the position of a particle by arranging the particle to collide with a screen. But if we had wanted to measure a different property – for example, the particle's momentum – we would have needed to apply a different operation.

So when we want to measure a quantum system, we first have to decide what property we want to measure, and then that decision determines the particular measurement operation which must be performed. Mathematically, we say we need to apply a different *operator* (an operator meaning a "mathematical operation" of some kind).

It turns out that the operator which has to be used in order to measure the energy of a system plays a particularly important role in quantum mechanics. In this section, we will be considering this energy operator, and discovering why it is so important.

Firstly, the concept of the *Hamiltonian* needs to be introduced. The Hamiltonian is named after the 19th century Irish mathematician William Rowan Hamilton. The Hamiltonian represents **the total amount of energy in a system.** It is a concept from classical physics which has carried-over into quantum mechanics. In equations, the Hamiltonian is conventionally denoted by the letter H.

The operator which is used to observe the amount of energy in a quantum system is therefore called the Hamiltonian operator, and it has the following form (I know this looks complicated, but don't worry – everything will be explained):

$$\hat{H}(\psi) = i\hbar \frac{\partial \Psi}{\partial t}$$

When the energy operator is written in this way, it becomes the most important equation in quantum mechanics – it is called the *Schrödinger equation.* You've probably heard of it already.

In fact, it was the discovery of this Schrödinger equation in 1925 which created the field of quantum mechanics. And, as we shall be seeing, it is also the key equation needed for building a quantum computer.

Let us consider the left-hand side of the equation first.

The left-hand side of the previous Schrödinger equation is just a statement that this represents the Hamiltonian operator (the total energy of a system), signified by the letter $\hat{H}$. The little caret symbol on top of the H simply indicates that this is an operator (in this case, the energy operator). Hence, it needs something to "operate" on. In this case, you can see that it is operating on the quantum state vector as the symbol (Ψ) is included immediately after the $\hat{H}$.

So all the left-hand side of the equation is doing is telling us that we are dealing with the energy operator operating on the quantum state vector. In other words, we are applying a mathematical operator to measure (or observe) the energy of a particular quantum system.

The right-hand side of the equation is more interesting, as this gives the actual form of the energy operator.

You will see on the right-hand side of the equation that there is a part of the equation which says: [2]

$$\frac{\partial \Psi}{\partial t}$$

This represents the amount by which the wavefunction, Ψ, changes with respect to time, t. Therefore, the Schrödinger equation reveals how we can change the wavefunction (quantum state) of the system over time: we just need to change the total energy of the system.

---

[2] This type of equation is called a *partial differential equation*. It means it considers the rate of change of a variable with respect to just one other, single variable – with all the other variables being held constant. In this case, it considers the rate of change of the wavefunction with respect to t.

Because of this, you might often read in quantum mechanics books that "the Hamiltonian determines the time evolution of the quantum state", or something similar. As an example, in their book *Quantum Mechanics: The Theoretical Minimum*, Leonard Susskind and Art Friedman say: "If we know the Hamiltonian, it tells us how the state of a system evolves with time". So this explains why the Hamiltonian plays such a vital role in quantum mechanics.

You will also see an $\hbar$ symbol on the right-hand side of the equation. Where does it come from, and what does it mean? Well, it's not too difficult. The $\hbar$ symbol (it is a letter h with a line through it, commonly called "h-bar") is the *reduced Planck constant*, equal to the Planck constant divided by 2π. The constant has to be included so that we are dealing with similar units on both sides of the equation. As an example, it would be meaningless if an equation was comparing apples with oranges, representing something like "five apples equals three oranges":

Comparing apples with oranges makes no sense. An equation only makes sense if it compares like-with-like, saying something like "five apples equals two apples plus three apples". The equation is then said to be "dimensionally correct":

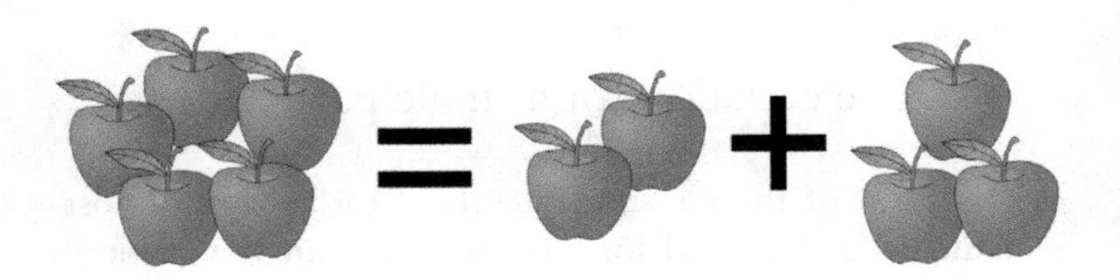

If you look at the Schrödinger equation, you will see that there is a term representing energy on the left-hand side of the equation, so the left-hand side of the equation has units of energy, but the

$$\frac{\partial \Psi}{\partial t}$$

term on the right-hand side of the equation represents the rate of change of the wavefunction with respect to time, the units of which are "inverse time", or "one divided by time". Therefore, it appears the Schrödinger equation is mistakenly comparing apples with oranges! The solution comes by multiplying the right-hand side by the reduced Planck constant, which has just the right units to convert the right-hand side of the equation into a value with units of energy. The resultant Schrödinger equation is then correctly comparing apples with apples, and has therefore become dimensionally correct.

This powerful type of analysis comparing the units on both sides of an equation is called *dimensional analysis.*

The equation for a single particle

The form of the equation presented earlier is the most fundamental and general form of the Schrödinger equation. However, you will often find the equation written in a slightly different form, aimed at describing a specific common example. That example is the Schrödinger equation for describing a single particle which is moving in a field of energy, for example, an electron moving between two electrically-charged metal plates. In that case, the total energy of the particle will have two terms: a kinetic energy term (kinetic energy is the energy associated with the particle's movement) and a potential energy term (the potential energy of the particle is the energy it receives purely due to its position in the field). Therefore, we can substitute these two amounts of energy into the "total energy" left-hand side of the Schrödinger equation. That then gives us:

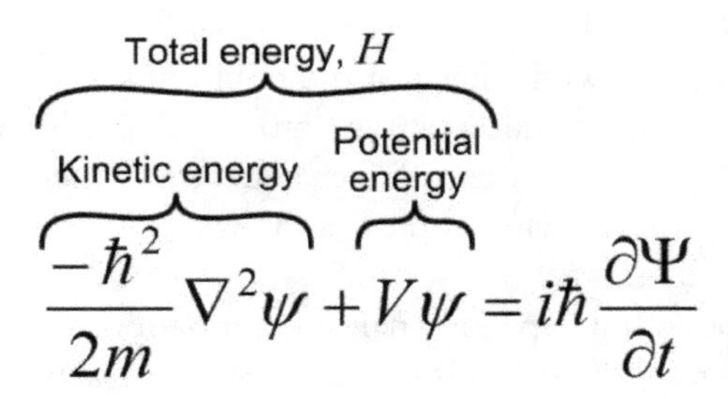

$$\frac{-\hbar^2}{2m}\nabla^2\psi + V\psi = i\hbar\frac{\partial\Psi}{\partial t}$$

You will see, though, that if you just replace the left-hand side of the equation (the sum of the kinetic and potential energy) with the total energy, *H*, you get back to the form of the Schrödinger equation presented earlier.

In their book *Quantum Mechanics: The Theoretical Minimum*, Leonard Susskind and Art Friedman describe this as the "iconic form of the Schrödinger equation that appears on T-shirts":

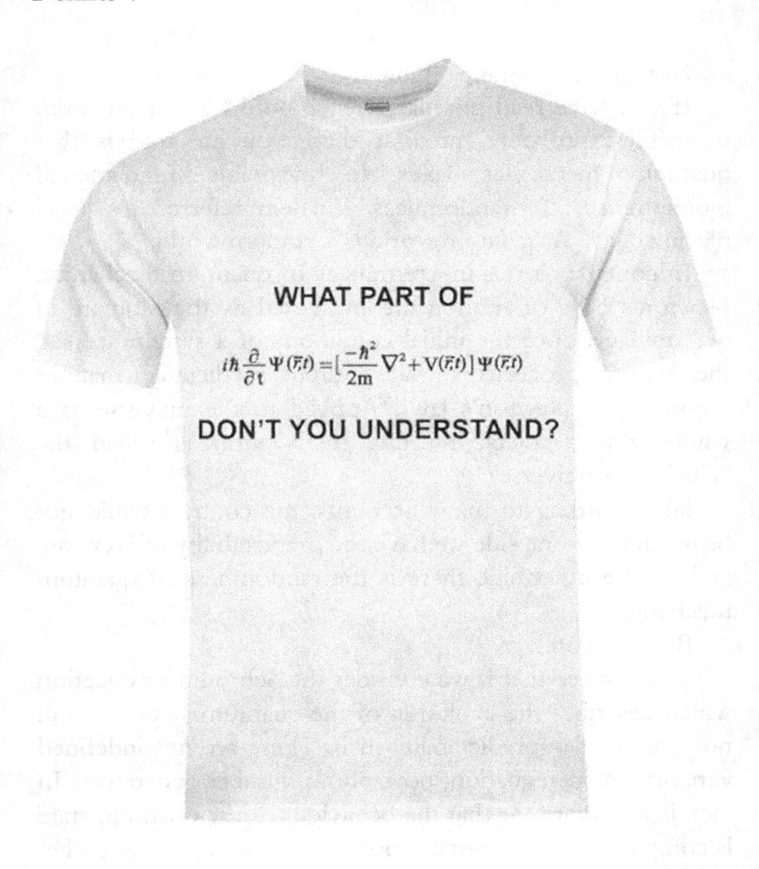

With reference to the logo on the T-shirt, hopefully you should now understand **all** of the parts of the equation on the T-shirt.

The complete predictability of quantum mechanics

Now, here's a strange thing.

If you have read popular science books about quantum mechanics, probably the first thing you are told is that quantum mechanics leads to inevitable fundamental indeterminacy, or randomness. Einstein referred to "God playing dice". A quantum world is a random world.

In contrast to this indeterminacy of quantum mechanics, Newton's laws of motion are presented as the ultimate in determinacy: once the initial conditions of a system are set, the system proceeds in an entirely predictable manner according to Newton's laws. Applied to the universe as a whole, this predictability due to Newton is called the "clockwork universe".

So, according to many accounts, the contrast could not be greater: on one side we have the predictability of Newton, and on the other side there is the randomness of quantum mechanics.

But, hold on.

You can see that if we consider the Schrödinger equation which describes the evolution of the quantum state, we find not a hint of unpredictability in it. There are no undefined variables in the equation, no random number generators. In fact, it would appear that the behaviour of the quantum state is completely deterministic – not random at all, and not what you might expect from "fundamentally random" quantum mechanics. Roger Penrose describes this surprising feature of the Schrödinger equation in his book *The Road To Reality*:

> *One thing that we note is that it is a deterministic equation (the time-evolution being completely fixed once the state is known at any one time). This may*

come as a surprise to some people, who may well have heard of "quantum uncertainty".

In fact, the equation appears to imply that the evolution of the quantum state is just as predictable as the evolution due to Newton's laws. So, what's up?

Well, it is completely true: the evolution of the quantum state as described by the Schrödinger equation is just as predictable and deterministic as Newton's laws. If you set the initial state, then the quantum state will evolve in an entirely deterministic manner – according to the equation. In fact, it seems like a perfect analogue to the determinism of Newton's laws.

So, once again, we find that the quantum world is not "weird" at all.

It is this determinism and predictability of the Schrödinger equation which allows the precise manipulation of the quantum state in quantum computers: if you change the energy, you know precisely how the quantum state will change. You have complete control.

It therefore appears that quantum behaviour is completely predictable. So why do hear about the "randomness" of quantum mechanics?

That is what we will discover in the next chapter …

4

OBSERVING A QUANTUM SYSTEM

In the previous chapter, the Schrödinger equation was introduced and it was explained how it describes the evolution of the quantum state. It was also explained how the Schrödinger equation was completely deterministic and predictable. So why do hear about the "randomness" of quantum mechanics?

Well, we have to remember that the quantum state vector we have considered so far represents the state of the system **before it is observed**, before we measure the system and obtain the value for some property of the system. As explained earlier, before observation, the quantum state can be in a mix of states: both red and blue at the same time. However, when we observe the system, we only ever observe a particle to be in a single, well-defined state: **either** red **or** blue – but not both. For example, in the double-slit experiment, we would only ever observe the particle to be in a single position – not in two places at once.

So, when we observe a quantum system, it is as if the state vector in a superposition state "collapses" to just a single value. This process has been called the "collapse of the wavefunction", and it is illustrated in the following diagram:

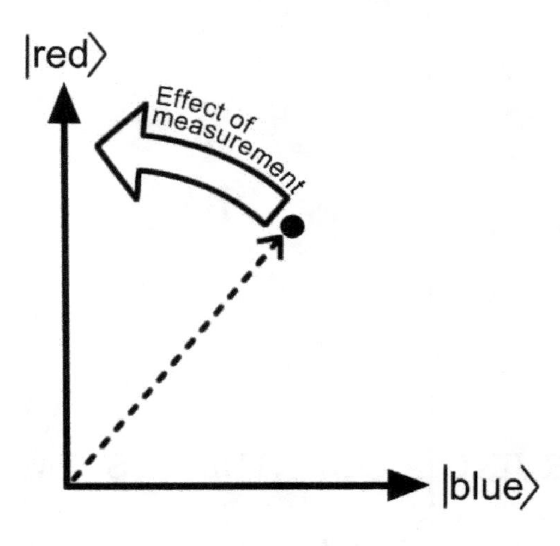

In the previous diagram, you can see what happens during the act of observation of a quantum system. You will obtain an observed value (either red or blue) but, as described earlier, the measurement operation also has the effect of modifying the state vector so that the vector moves from its superposition state to point at the value which has been observed (either red or blue).

So there is a rotation of the state vector from its superposition state to point to a well-defined single state.

Some new terminology needs to be introduced at this point. Firstly, the "well-defined" states – the states which we can possibly observe when we make a measurement – are called *eigenstates*. Secondly, the state vector which represents a particular eigenstate is called an *eigenvector*. So, when we make a measurement, the state vector rotates to become an eigenvector, and the system is then said to be in an eigenstate.

Now we get to the crucial question: what determines which of the possible states we will observe when we measure the system? As an example, in the previous diagram, it seems that the state vector could just as easily have rotated to become a blue state. In that case, why was red selected?

The answer is that the observed value is determined by the *projection* of the state vector onto the relevant eigenvector. This is perhaps easiest explained by the following diagram:

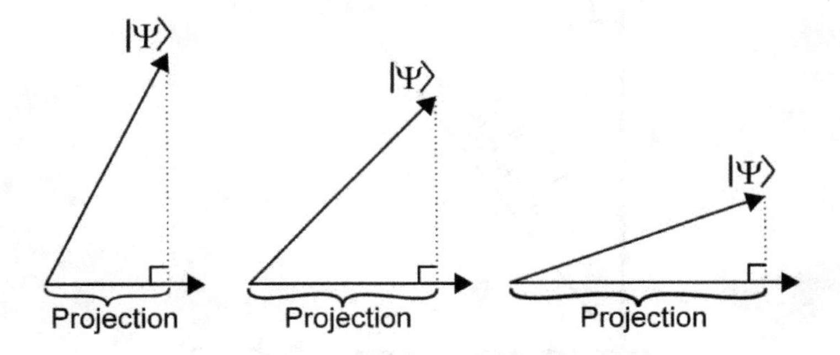

You can see from the previous diagram that – to calculate the projection – a dotted line is first drawn from the tip of the state vector to make a right angle with the eigenvector under consideration (in this particular diagram, the eigenvector is the horizontal line). The projection is then defined as the distance along the eigenvector (along the bottom horizontal) until you reach the dotted line. This is all clearly shown on the previous diagram.

Crucially, the value of the projection then represents the **probability** of that particular eigenstate being selected. **So it is at this observation stage that probability enters quantum mechanics.**

Let us consider an example of how to calculate the value of the projection. Consider the following diagram. A state vector is shown with the tip of the vector at the coordinates (0.6, 0.8):

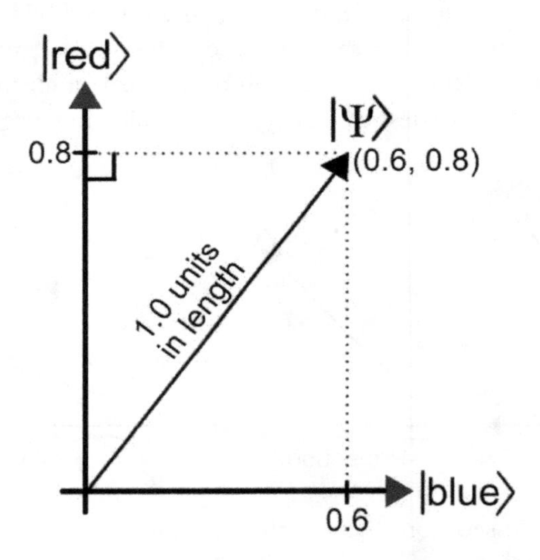

The vector is 1.0 units in length, in which case it is called a *unit vector* (state vectors are usually defined as unit vectors).

We would like to know what value we will obtain when we observe this particular quantum system. Will the state vector rotate so that it points in the vertical direction (towards red), or will it rotate so that it points in the horizontal direction (towards blue)? In other words, will we observe the particle described by this quantum state vector to be either red or blue?

Well, as stated earlier, the selection of the observed value is determined by the projection of the state vector onto the relevant eigenvector. From the diagram, it is easy to see the relevant projection values from the coordinates of the state vector. You can see the projection along the red eigenvector is 0.8 units, and the projection along the blue eigenvector is 0.6 units.

Remember, these projection values merely give you the **probabilities** of a particular value being obtained after observation. However, in that case, then these particular values of 0.8 and 0.6 cannot be quite correct. We see that there are only two possible outcomes from the observation: **either** we observe red, **or** we observe blue. With only two possible outcomes, we know that the sum of the probabilities of these two outcomes must be 1.0 (as probabilities always sum to 1.0). So, to calculate the correct probabilities, we have to take the **square** of the projection values, which gives us 0.64 (which is 0.8×0.8) and 0.36 (which is 0.6×0.6). You can see that these new values do, indeed, sum to 1.0. So these are the correct probabilities: the probability of observing red will be 0.64, and the probability of observing blue will be 0.36.

If you consider the previous diagram, you can easily see why this "squaring" rule will always lead to the sum of the probabilities being equal to 1.0. It is due to Pythagoras's theorem: the length of the state vector has been set to 1.0. The state vector is on the hypotenuse of a right-angled triangle, so the sum of the squares of the other two sides of the triangle (the sum of the squares of the two projection values) must therefore be equal to 1.0.[3]

So it is during the act of observation that probabilities enter quantum mechanics, and that is the reason why quantum mechanics has obtained its reputation for randomness and indeterminism.

---

[3] This provides the explanation for the *Born rule*, which states that the probability of finding a particle in a particular location is proportional to the **square** of the particle's wavefunction at that point.

Calculating eigenvectors and eigenvalues

As far as quantum computing is concerned, this chapter has now explained everything you need to know about what happens when a quantum system is measured. So, if you wish, you may now skip to the next chapter. However, in the remainder of this chapter we will go into a bit more detail about how the eigenstates are calculated, and how those eigenstates explain the behaviour of electrons as they orbit in an atom.

However, like I say, if you do not want to know these details then feel free to skip to the next chapter.

In the previous section it was explained how when we observe (or measure) a quantum system we will only ever observe it to be in a single, well-defined eigenstate, with clearly-defined values. For example, we might find a system in either the red or blue state. However, no explanation was provided as to how those eigenstates – and their associated property values – could be calculated. Where do they come from?

It was stated in the previous chapter that observed values are produced by mathematical operators, acting on the quantum state vector. But, with just a little thought, it can be realised that the application of the operator cannot be entirely straightforward.

To understand why that is the case, it must be realised that the quantum state vector might be pointing in any arbitrary direction before observation. Simply applying the operator to the state vector would then produce a completely arbitrary result – some mixture of red and blue, for example – whereas we need a clearly-defined result from the operation: **either** red **or** blue. Some arbitrary mixture is not good enough.

So the operator must be used in applied in rather a different mathematical manner. But how should we apply it? Well, we have a clue …

Imagine that the state vector before observation already points in the direction of a well-defined state. For example, it might be pointing in the red direction or in the blue direction. There are two important points which arise in that case.

Firstly, if the state vector is already perfectly aligned with, say, the red eigenvector then the projection of the state vector onto that eigenvector will then be equal to 1.0, meaning they are pointing in precisely the same direction. That means the probability of measuring red will be 1.0. In other words, there will be no uncertainty at all in the measurement (a probability of 1.0 means certainty).

The second point is that if the operator applied to that state vector, it can be realised that **the state vector will not be modified.**[4] The state vector will not be changed because it is already pointing to a clearly-defined state – either red or blue – so there is no need for the state vector to be rotated at all.

So this is a big clue, and it gives us all the information we need. As can be seen from the following diagram, if the measurement operator is applied to a well-defined state vector (for example, red), the result will be an unchanged

[4] As James Cresser of Macquarie University has said: "It can also be argued that if, after performing a measurement that yields a particular result, we immediately repeat the measurement, it is reasonable to expect that there is a 100% chance that the same result be regained, which tells us that the system must have been in the associated eigenstate."
http://tinyurl.com/jamescresser

state vector (still pointing to red). Actually, as only the direction of the state vector is important to define a quantum state, we can say that the result can be some multiple of the original state vector:

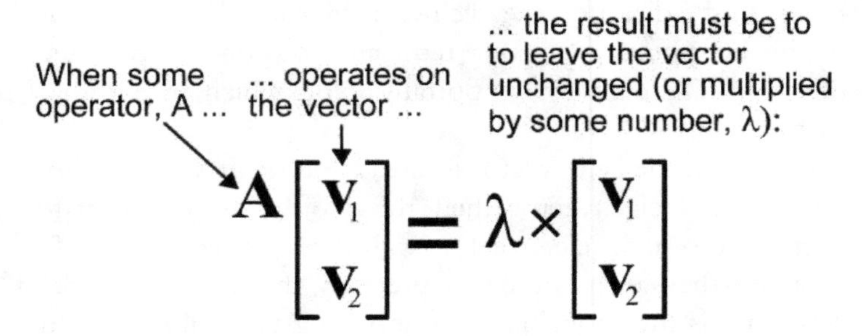

This type of relationship is well-known in mathematics. Vectors which satisfy this relationship are called *eigenvectors* (which were introduced in the previous section). It turns out that if we have a particular operator (such as the operator A in the previous diagram) there will only be a few eigenvectors which satisfy the relationship. We might consider these as representing the allowed state vectors (for example, the state vectors red and blue). So this explains why, when we measure a quantum system, we only observe particular values (red or blue, for example). It is because only certain eigenvectors will satisfy the previous relationship.

The quantum state which is associated with a particular eigenvector (the state of red or the state of blue, for example) is called an *eigenstate* (again, a term which was introduced in the previous section).

There is something else interesting in the previous diagram. If a measurement is made, it so happens that the value of the property you will measure will be equal to the λ multiplying factor shown on the previous diagram. The value of λ – the property value you will observe – is then called an *eigenvalue*.

And that is how you find the eigenvectors and eigenvalues (red, blue, etc.) associated with a particular measurement operator. That is how you discover the discrete states and values a quantum system will take after it is observed.

Energy levels of an electron

An atom is composed of an atomic nucleus (made of positively-charged protons and electrically-neutral neutrons) which is orbited by electrons. The "height" – and corresponding energy levels – of those orbiting electrons represents one of the most important examples of eigenstates and eigenvalues.

The puzzle over the energy levels of electrons was one of the great puzzles of the "old quantum theory" period from 1900-1925. It was known that when a substance was heated, it would glow ("red hot" or "white hot") and the emitted radiation represented a form of energy loss from the atom. In 1900, the German physicist Max Planck realised that the radiation being transmitted from the atoms took the form of chunks of energy (known as "quanta"), a realisation that started the "old quantum theory". In 1905, Albert Einstein correctly proposed that the emitted quanta of energy were actually particles, particles which we now call photons. Photons are the particles of light, the particles which are emitted when a substance glows red hot or white hot.

The energy of the emitted photons can only take particular values – not continuous values. This mystery can be understood by considering the energy levels of electrons which are orbiting an atomic nucleus – and then applying our new-found knowledge of the Schrödinger equation (remember: the Schrödinger equation can be used to describe the energy of a particle).

If we consider a diagram of an atom, with the nucleus at the centre surrounded by orbiting electrons, we find that only particular electron orbits are allowed around the nucleus:

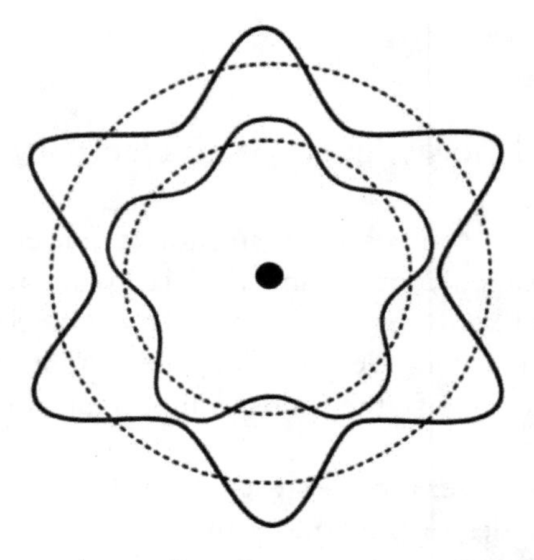

On the previous diagram you can see two different electron orbits, shown by the two wavy lines. Only integer values (whole numbers) of waves are allowed around the orbit. You can see that the inner orbit has five complete waves, whereas the outer orbit has six complete waves. These orbits are allowed, but electrons are not allowed to orbit in the spaces between these orbits.

The different orbits represent different energy levels of an electron. Remember from the previous chapter that the Schrödinger equation describes the energy of a particle. The orbits of an electron can then be calculated as the eigenstates of the Schrödinger equation of that electron. Considering the geometric picture of the atom, we can see that the allowed orbits of the electron would represent eigenstates of energy of the electron (with only a few particular eigenstates being allowed). The associated energy levels of the electron would then be the eigenvalues.

One of the great mysteries of the "old quantum theory" in the first two decades of the 20th century was why photons are only ever released from atoms with certain discrete chunks of energy. However, in 1925, with the discovery of the Schrödinger equation, the mystery could finally be solved. Remember, in the atom, eigenstates of energy have discrete values – they are not continuous (capable of taking any value). So when an electron "jumps" from a higher energy eigenstate to a lower energy eigenstate, a photon is released which has energy equal to the difference between the two energy states. This explains why photons are only ever released with certain discrete chunks of energy.

So the theory of quantum mechanics was able to explain the structure of the orbits of electrons in an atom, and the origin of the "quanta" of energy. One the great mysteries of the "old quantum theory" could now be clearly and logically explained by the new quantum mechanics – with the Schrödinger equation at its core.

It also explains why it is best not to treat the development of the theory of quantum mechanics in chronological order: quantization arises naturally when the Schrödinger equation is discovered first – not vice versa.

5

COMPLEX NUMBERS

The use of *complex numbers* is widespread in many areas of science, engineering, and mathematics. In quantum mechanics, complex numbers play a central role.

You might already have an understanding of complex numbers, but a short explanation is that a complex number is a number which is composed of two parts which are added together.

Firstly, the complex number has a so-called *real* part, which is just a number in the usual, conventional sense (for example, 5, 3000, or 12732.506). Secondly, the complex number has a so-called *imaginary* part, which is a number multiplied by i, where i represents the square root of -1. You might find it strange to consider any number as representing the square root of a negative number, after all, when any number – positive or negative – is squared, that will always result in a positive number. So the fact that i cannot represent a conventional number explains why it forms the so-called "imaginary" part of a complex number.

Here is an example of a complex number:

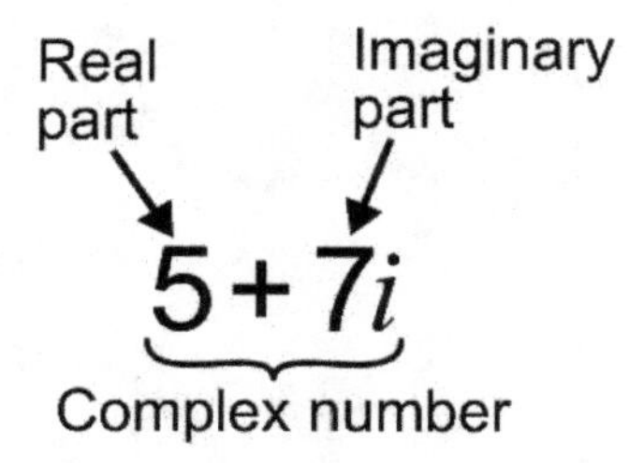

In the chapters so far, we have seen that the quantum state vector exists in a "vector space", and it can rotate in that vector space. The particular type of vector space which is used in quantum mechanics is called a *Hilbert space* (named after the great German mathematician David Hilbert, who we have encountered in my previous books).

The important feature of a Hilbert space is that the coordinates of points can be complex numbers. Therefore, the coordinates describing quantum state vectors can be complex numbers.

The phase of Hilbert space

Let us now consider how the complex nature of the state vector gives each state vector a *phase angle* value.

If you remember, a complex number is composed of two parts: a real part and an imaginary part. Let us consider a complex number with a magnitude of 1.0. In other words, if you were to square the real part of the number, and add it to the square of the imaginary part of the number, the result would be 1.0. From Pythagoras's theorem, we could plot the resulting number on a circle, with a radius of 1.0 (the complex number being a distance of 1.0 from the centre of the circle), as shown on the following diagram:

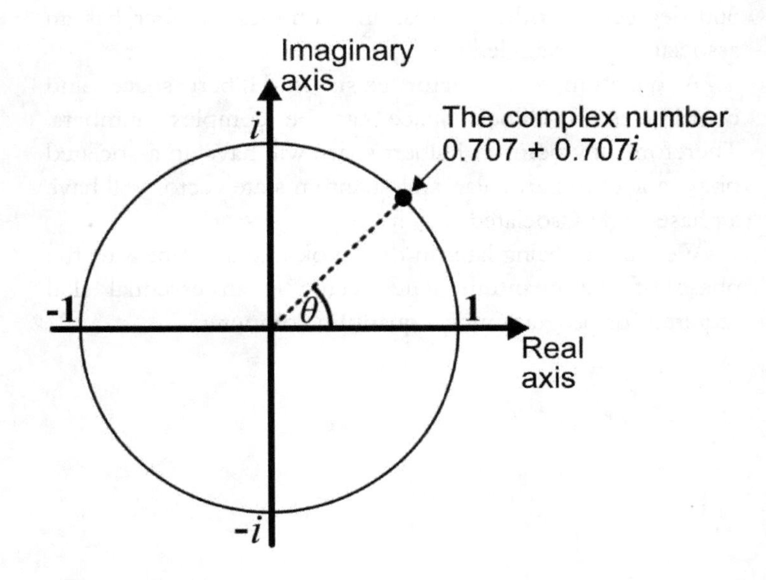

You can see from the previous diagram that the real part of the number is plotted along the horizontal axis, while the imaginary part of the number is plotted along the vertical axis. You will see that one particular complex number has been plotted as a point on the circle ($0.707+0.707i$, where 0.707 is the square root of 0.5, so, by Pythagoras's theorem, the complex number has a magnitude of 1.0 and can be plotted on the circle).

This method allows us to think about complex numbers in geometric terms, the complex number being represented by a point at the end of a line from the centre of the circle.

You can also see from the previous diagram that the line from the centre of the circle to the complex number is at an angle θ around the circle measured in an anticlockwise direction from the horizontal axis. We can think of this angle as a *phase*, the phase being able to take any value from 0 to 360 degrees. In other words, any complex number has an associated phase angle.

A quantum state vector exists in Hilbert space, and coordinates in Hilbert space can be complex numbers. Therefore, any point in Hilbert space will have an associated phase angle. In particular, any quantum state vector will have a phase angle associated with it.

We will be seeing later in this book that awareness of the phase of the quantum state vector is an essential skill required for programming a quantum computer.

6

DECOHERENCE

Quantum mechanics was developed in the early decades of the 20th century. This was a period of global revolution, when the old rules and certainties were discarded. These developments were also reflected in the modernist art of the period, with artists influenced by the new developments from physics.

As an example, artists such as Picasso who pioneered the Cubism movement moved away from the rigid rules of three-dimensional perspective. The rules of perspective implied that the observer was positioned at a single point in space, as shown in the following line drawing by the 16th century Dutch renaissance artist Hans Vredeman de Vries:

The rules of perspective implied that the observer was irrelevant, that the reality of an object or scene could be captured in its entirety even if the position of the observer was fixed in one place.

With the arrival of quantum mechanics in the early 20th century, however, it was realised that the observer was important, that reality was generated from a combination of the observer and the subject.

This principle influenced the Cubism movement, as seen in the following *Portrait of Dora Maar* by Picasso (1937). Picasso attempts to capture Dora Maar from several viewpoints – several observers – and combine those into a single portrait. You can see that the result is a front view of Dora Maar, in combination with a smaller view from the side.

In essence, the following portrait might be considered to be an attempt to represent Dora Maar in a quantum superposition state:

Modernist artists were also influenced by the recent discovery of relativity and the philosophy of Henri Bergson, both of which suggested that time should be considered as being the fourth dimension. Here is an example of Picasso playing with spatial dimensions in his 1910 painting *Girl with a Mandolin*, just four years after Einstein discovered relativity. Picasso is referencing the fourth dimension by stacking a succession of time-slices, "sticking together several three-dimensional spaces in a row":[5]

---

[5] *The Birth of Cubism*, **http://tinyurl.com/birthofcubism**

So art in that era was reflected by developments in physics. But sometimes the relationship works the other way round – sometimes art discovers fundamental truths before physics. This was true in the case of another early 20th century artistic movement: Futurism.

The Futurist movement was based in Italy. The Futurists believed that it was wrong to consider objects in isolation, and that the surrounding environment always played a role in defining the object. As an example, here is a photograph of Umberto Boccioni's 1913 bronze sculpture *Unique Forms of Continuity in Space*. The sculpture depicts a human form in fluid motion, which is blurring into the surrounding space.

The result is that the boundary where the human ends and the surrounding environment begins has been blurred. The sculpture can be seen on the reverse of the current Italian twenty cent coin:

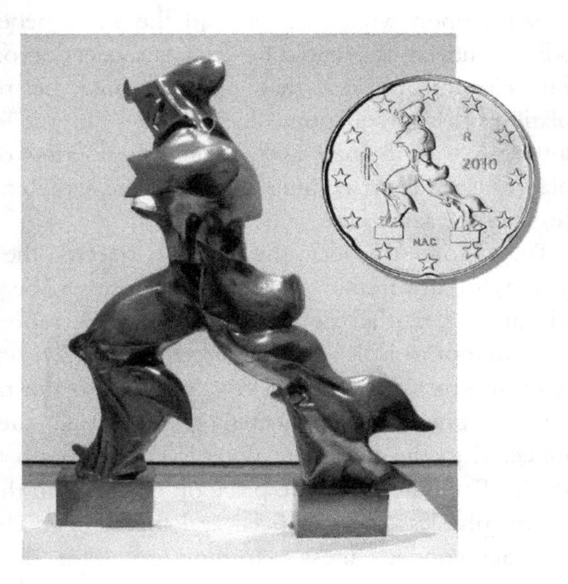

The Futurists took their lead from the writings of the philosopher Henri Bergson:

> *Does not the fiction of an isolated object imply an absurdity, since this object borrows its physical properties from the relations which it maintains with all the others, and owes each of its determinations, and consequently its very existence, to the place it occupies in the universe as a whole?* ***Any division of matter into independent bodies with determined outlines is artificial.***

I have placed the last sentence in emphasis: "Any division of matter into independent bodies with determined outlines is artificial". In 1911, the Futurists released their manifesto describing their philosophy about art: "Our bodies penetrate the sofas upon which we sit, and the sofas penetrate our bodies." Indeed, this could be seen as a rejection of the idea from classical physics that objects could be treated in isolation. Just as quantum mechanics – in the early 20th century – was moving away from considering objects in isolation, so was art coming to exactly the same conclusions at exactly the same time.

To reinforce this idea, the Futurists rejected the so-called "tyranny of the frame", preferring pictures to be displayed without frames, the frame around a picture representing a futile attempt to isolate the picture from its environment. In reality, the picture is physically connected to the frame, the frame is physically connected to the wall, the wall is connected to the floor, the floor is connected to the street outside. The boundaries we place on objects – in the style of classical physics – are artificial and misleading. There is an inevitable connectedness. Nothing can ever be perfectly isolated.

Except for, perhaps, one exception …

… the dark, cold, interior of a quantum computer.

Environmental decoherence

In 1970, Dieter Zeh of the University of Heidelberg published a paper which suggested a mechanism which could explain the mysterious "collapse of the wavefunction".[6] To understand Dieter Zeh's idea, we need to return to Chapter Two of this book in which it was explained that quantum mechanics inevitably arises when we take a more realistic and sophisticated picture of how the world must work.

In Chapter Two, two fundamental principles were presented:

- A measurement inevitably affects the object under observation. We are therefore not measuring the true reality of the object on its own, i.e., the object before it is measured.
- Objects are never completely isolated. The measurement we eventually obtain is a measurement of the joint system: "An electron being measured by a measuring device".

It is the second of these two principles which formed the basis for Dieter Zeh's explanation of the apparent "collapse of the wavefunction". It is the principle that objects are never completely isolated from the rest of the universe. The idea that our world is split into entirely discrete, isolated objects is a completely human invention. Think of yourself

---

[6] *On the Interpretation of Measurement in Quantum Theory*, H.D. Zeh, **http://tinyurl.com/zehpaper**

sitting in your chair. You probably think of yourself as being a separate object from the chair. But the atoms of your backside are pressed against the atoms of the chair, in constant contact. So where is the division? Does that not represent one joined unit? And we could continue the process: your chair is in direct contact with the floor. The floor is directly connected to the Earth. Again, where is the division? How can you justify thinking that you are an entirely separate object from the chair or the planet?

As another example, you breathe air. The air is part of a continuous substance which surrounds the planet and places you in contact with every other living person. In that case, how can you consider yourself to be an entirely separate object?

Of course, you should not think of yourself as being isolated. As has just been described, this was the belief of the Futurist modern art movement. And it is this principle which led Dieter Zeh to propose a mechanism behind the apparent "collapse of the wavefunction". In particular, Dieter Zeh suggested that it was the ever-present environment surrounding an object – constantly "observing" the object – which leads to the object being taken out of its superposition state.

The mechanism is called *decoherence*.

And, once again, we encounter the idea that quantum mechanics is more of a "state of mind", representing a more sophisticated picture of how the world must work. In this case, the principle is that no object is ever completely isolated from the rest of the world.

Unfortunately, Dieter Zeh passed away earlier this year. He should surely have received the Nobel Prize in his lifetime for his work in proposing decoherence, which has now been firmly experimentally verified.

We can use a simple thought experiment to explain how decoherence works, and the thought experiment we will use is the double-slit experiment which was described in Chapter Two.

In the double-slit experiment, a beam of light is directed towards two narrow slits, with a screen positioned on the other side of the slits. The light beam passes through both of the slits.

If, at the screen, a peak in the wave from one slit coincides with a peak in the wave from the other slit, then a bright point appears on the screen. Conversely, if a peak in the wave from one slit coincides with a trough in the wave from the other slit, the two waves will cancel at that point and no bright point will appear on the screen. Therefore, as shown in the following diagram, a regular pattern of dark and bright bands will appear on the screen:

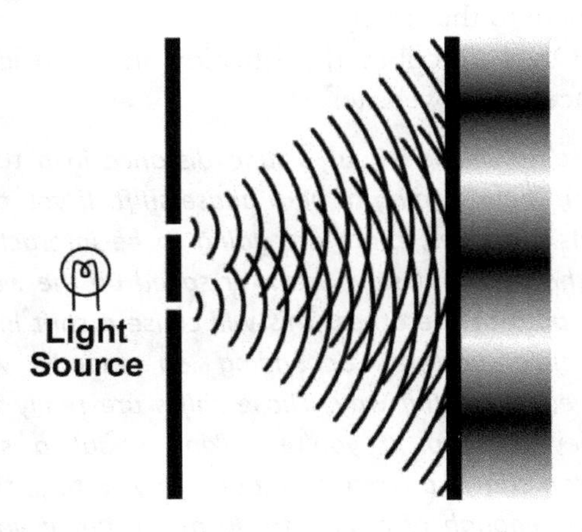

The light intensity is then reduced so that only a single photon of light is in transit at a time. Over time, a pattern develops on the screen as the individual photons strike the screen and accumulate. Amazingly, though, the interference pattern is still produced – even though there is only one photon in the system at a time! It is as if the single photon passes through both slits and interferes with itself!

Let us now make the experiment rather more realistic. We must realise that particles, in general, are never completely isolated from their environment, and that environment will tend to be rather messy and random. Let us consider the effect of the messy, random environment on the single photon.

In a more realistic picture, on its way from the light source to the screen, the photon has to pass through a considerable distance of air (representing the "environment"). That air is likely to be turbulent, moving randomly. As it moves along its path, the photon interacts with the atmosphere, and it, in return, is affected by the atmosphere. The effect of the random, turbulent atmosphere will be to slightly increase or decrease the length of the particle's path to the screen.

Chad Orzel describes the situation in an article about decoherence on his website:[7]

> *If we're talking about a long distance in a turbid medium, there's going to be a phase shift. If you think in terms of waves, there are going to be interactions along the way that slow down or speed up the waves on one path or the other. This will cause a shift in the interference pattern, depending on exactly what happened along the way. Those shifts are really tiny, but they add up. If you're talking about a short interferometer in a controlled laboratory setting, there won't be enough of a shift to do much, but if you're talking about a really long interferometer, passing*

---

[7] *Many-Worlds and Decoherence: There Are No Other Universes*, Chad Orzel, **http://tinyurl.com/orzeldecoherence**

through many kilometers of atmosphere, it'll build up to something pretty significant.

The effect of this shortening and lengthening of the path is to very slightly move the position of the peaks and troughs of the light wave. Because the positions become random, the effect is to destroy the interference pattern on the screen.

And, if the interference pattern is destroyed on the screen, that makes it appear as though the photon only went through one slit. Remember, the interference pattern is caused by a photon apparently going through two slits at once. So if the interference pattern is destroyed then the photon no longer appears to be in a superposition state, no longer going through two slits at once.

So we can treat the interference pattern as being an indirect signature of the photon being in a superposition state. And the presence or absence of the interference pattern indicates whether or not the particle is in a superposition state.

At this point, we can put our Sherlock Holmes deerstalker hat on and deduce what is responsible for the apparent "collapse of the wavefunction", taking the particle out of its superposition.

The initial arrangement of the double-slit experiment produced an interference pattern on the screen. From that interference pattern, we deduce that the particle is in a superposition state.

The second form of the experiment was more realistic, introducing a random environment. The result was that the interference pattern was destroyed. We therefore deduce that the particle is not longer in a superposition state.

So what was responsible for taking the particle out of its superposition state? In other words, what was responsible for the apparent "collapse of the wavefunction" which seems to have occurred?

Well, with our Sherlock Holmes hats on, we observe that the **only** factor that has changed in the two versions of the experiment has been the introduction of a noisy, messy, random environment. Sherlock Holmes once famously said: "when you have eliminated the impossible, whatever remains, however improbable, must be the truth". On that basis, we have all the information we need to make our final deduction:

The apparent "collapse of the wavefunction" is caused by a particle interacting with the noisy, messy, random environment.

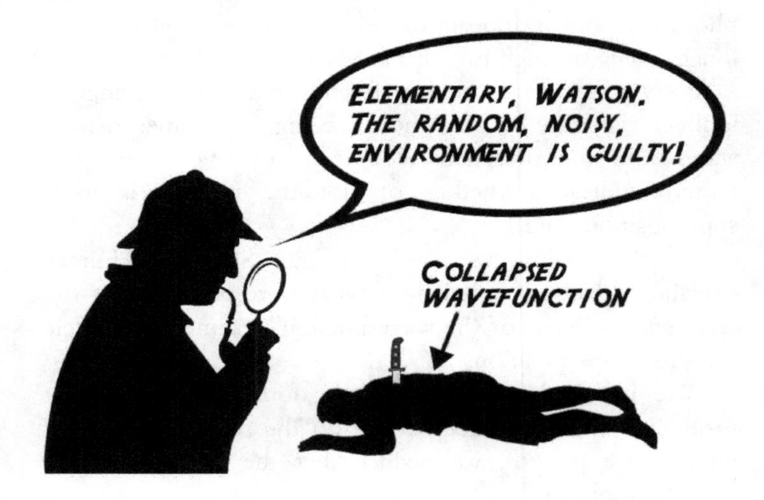

And that is the principle behind decoherence.

But what is actually going on at the microscopic level? How is the random environment capable of taking a single particle out of its superposition state? Well, if all we are interested in is building a quantum computer, or programming a quantum computer, we really don't care. All we have to do is make sure we isolate our quantum computer from its surrounding environment in order to maintain particles in their superposition state. And it is quite fortunate that we don't need to know precisely what is happening at the quantum level because the science of decoherence is not yet fully understood. [8]

However, we do have a fair idea of the sort of process which must be occurring at the microscopic level. Let us consider a particle hitting a screen (with the screen representing the "environment"). It must be realised that the screen is already in a well-defined, classical state. In other words, the screen is not in a superposition state – it is in an eigenstate, meaning the screen only exists in one position. When the particle hits the screen (interacts with the environment), the state vector of the particle finds itself in competition with the state vector of the much larger environment. This means the state vector of the particle – in its superposition state – is going to be affected more than the state vector of the screen.

There will be components of the particle state vector which do not agree with the state vector of the screen. Those components disperse into the wider, noisy environment, much like the splash from a stone thrown into a rough, turbulent ocean. The Schrödinger equation continues to apply to all those components, controlling their evolution in time. In other words, there is no new sudden mystical "collapse" process. However, once those components have dispersed into the "ocean", it becomes effectively impossible to regenerate the original superposition state (this explains

---

[8] The clearest technical descriptions of decoherence I have found are by John Boccio, formerly of Swarthmore College:
http://tinyurl.com/johnboccio
http://tinyurl.com/johnboccio2

why the apparent "collapse of the wavefunction" is an irreversible process). Those superposition elements – to all intents and purposes – disappear. The particle is then detected in only one position: on the screen.

In his book, *The Fabric of the Cosmos*, Brian Greene describes the process: "Decoherence forces much of the weirdness of quantum physics to 'leak' from large objects since, bit by bit, the quantum weirdness is carried away by the innumerable impinging particles from the environment."

So it is almost impossible to completely isolate an object from the rest of the world. But if we want to maintain a superposition state inside our quantum computer, we need to find some way of achieving that isolation, and thereby preventing that onset of decoherence. Let us now examine how we can achieve that isolation.

The coldest place in the universe

As we have just seen, any form of randomness in the environment can result in decoherence. In particular, any heat in the environment will result in the random motion (thermal noise) of the molecules in the environment. The resultant buffeting of a single particle by those random molecules will result in decoherence, taking the particle out of its superposition state. We will be seeing that quantum computers rely on particles being kept in superposition states for as long as possible. For this reason, quantum computers must be kept at extremely low temperatures to reduce thermal noise to a minimum.

The core of a computer must be cooled to just a tiny fraction of a degree above absolute zero – 180 times colder than deep space. Incredibly, this means that **the core of a quantum computer is the coldest place in the entire universe!**

The following photograph shows the internal structure of an actual IBM quantum computer. You can see that it looks very beautiful and intricate – it has been affectionately described as a "steampunk chandelier". The quantum processor is located in the metal cylinder at the bottom of the construction. The metal cylinder acts as a shield to protect the quantum processor from electromagnetic radiation (which could cause decoherence):

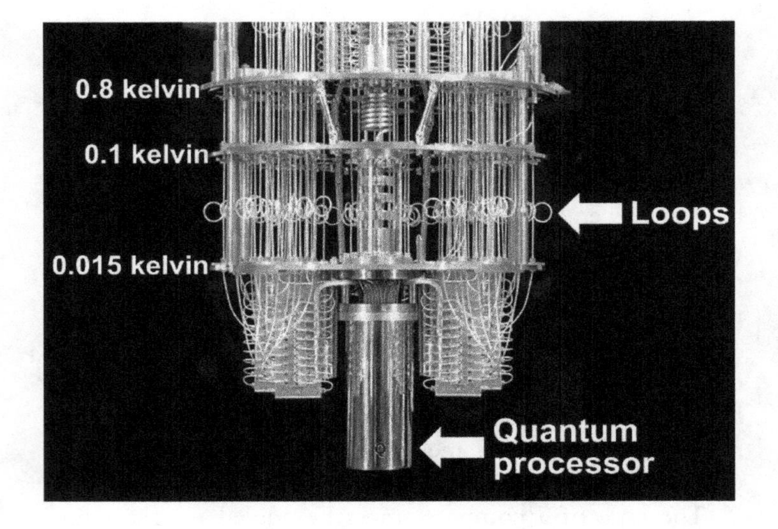

The whole device has to be maintained at an extremely low temperature. However, the signals transmitted down the cables to the quantum chip originate from control units outside the quantum computer – which are therefore at room temperature. So those cables have to go on a journey from room temperature to a temperature colder than space as they travel down the quantum computer structure. As you can see on the previous diagram, by the time the cables reach the top of the device, the temperature has been reduced to about 0.8 kelvin. You can see on the diagram that there are a series of horizontal plates, with the temperature being progressively reduced on each plate. The temperature continues to decrease toward the base of the device, with the actual processor chip being kept at a temperature of just 0.015 kelvin (15 milli-kelvin). At such a low temperature – just a fraction of a degree above absolute zero – atoms become virtually motionless.

The cables are made out of a superconducting material to minimise energy loss (when superconducting materials are cooled to near absolute zero, electric currents can pass through them without experiencing any electrical resistance).

On the previous diagram, you will also notice some ornate loops. These seem to add to the beautiful "chandelier" effect, but they most certainly serve a purpose. The loops are in the coaxial cables which transfer the data signals to-and-from the quantum processor. As the cables contract due to the low temperature, their length decreases and they could break. The loops in the cables ensures there is always some excess length in the cables to prevent them breaking. You can see similar loops on telegraph poles to stop the cables breaking in cold weather.

The whole device is then contained within a larger cylinder, which acts as a refrigerator, as shown in this photograph which was taken in the IBM Research Center in Yorktown Heights, New York:

The race against time

It is clearly extraordinarily difficult to maintain a particle in its superposition state. The key strategy is to keep the particle isolated from the random, turbulent environment. But even a single stray photon striking an electron effectively performs a "observation" on the position of the electron, and that will be sufficient to take the electron from its superposition state. It gets even worse than that. We might make efforts to eliminate all electromagnetic radiation (photons) from our environment, but there is no shield for gravity. Any change in the gravitational field – the movement of the Moon or Sun, for example – will disturb the particle.

This effect was quantified by John Boccio.[9] If we have a mass, m, which changes its height by a distance h, then the change in its gravitational energy will be equal to mgh (where g is the acceleration due to gravity).

Then, after a time, t, which is equal to:

$$t = \frac{\hbar}{mgh}$$

the change in energy will be sufficient to induce decoherence in the mass. If the calculation is performed for a mass of 100 kilograms with h only the length of a single atom, then this decoherence will occur in the extraordinarily short time of 10^{-27} seconds!

[9] **http://tinyurl.com/johnboccio**

This incredibly short time window can be extended by improving the isolation from the environment.

Later in this book, you will be using the IBM Q Experience website to program a quantum computer. On the web page, you will be able to see the real-time details of each quantum processor just be clicking on its name, as shown in the following diagram:

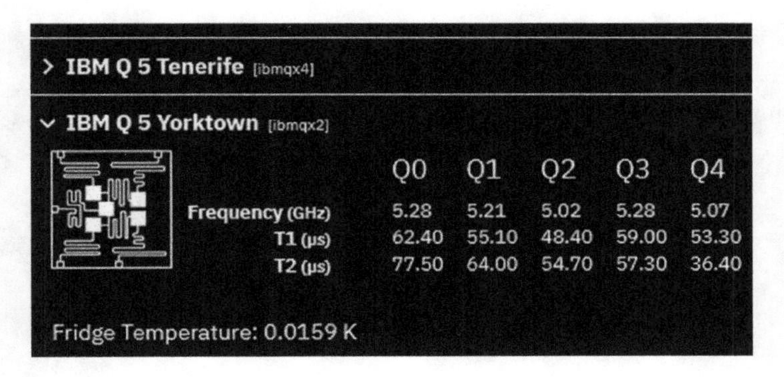

	Q0	Q1	Q2	Q3	Q4
Frequency (GHz)	5.28	5.21	5.02	5.28	5.07
T1 (µs)	62.40	55.10	48.40	59.00	53.30
T2 (µs)	77.50	64.00	54.70	57.30	36.40

On the previous diagram, the T2 "coherence times" measure how long it takes for decoherence to destroy the superposition state. For the five quantum bits (or "qubits" – see next chapter) on this particular processor, you can see that the T2 times vary from 36.40 microseconds to 77.50 microseconds. To put that into perspective, maintaining a quantum superposition for that length of time is at the top-end of what is currently possible using today's technology. Also note on the diagram that the "Fridge Temperature" is currently being measured at 15.9 millikelvins – just a fraction of a degree above absolute zero.

Any calculation has to be performed while the particle can be kept in its superposition state – the entire calculation must be performed before coherence is lost. As we saw in the previous diagram, that represents a time period of just a few microseconds. It is therefore clear that, for a quantum computer, any calculation is a race against time.

7

THE QUBIT

In my previous book it was explained how information can be measured in terms of *bits*. A bit (short for "binary digit") is a unit of data which can be either 0 or 1. In a conventional computer, all data is composed of combinations of these bits. In the computer, those bits are represented by different electric voltages: often +5 volts representing a 1, and zero volts representing a 0.

In a quantum computer, the equivalent of a bit is a *qubit* (short for "quantum bit"). In a qubit, the 0 and 1 values represent quantum states, so they are denoted by $|0\rangle$ and $|1\rangle$ (note the use of the Dirac notation). In a quantum computer, all processing occurs before decoherence destroys the superposition state of the qubit. That means a qubit can be in a superposition state and can have the value of **both $|0\rangle$ and $|1\rangle$ at the same time.**

However, when it is observed, it will be found in one of two possible states, $|0\rangle$ or $|1\rangle$.

The following diagram shows a graphical representation of a qubit in a superposition state:

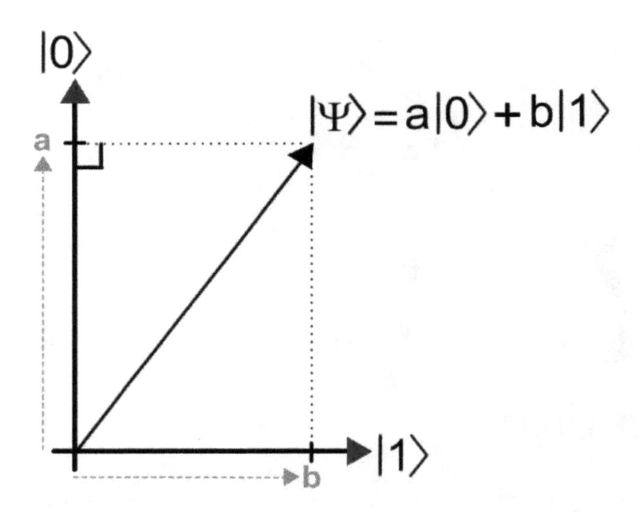

As shown on the previous diagram, the state vector of a qubit can be described by the following:

$$\psi = a \mid 0\rangle + b \mid 1\rangle$$

where a and b are complex numbers (because the state vector is in Hilbert space).

The physical implementation of qubits

If we are only interested in programming a quantum computer, then the actual physical implementation of a qubit on that computer is not of particular interest to us. As Alexandre Zagoskin says in his book *Quantum Mechanics: A Complete Introduction*: "In principle, any quantum system with a two-dimensional Hilbert space can be a qubit". A qubit can be physically implemented in several different ways, but as long as it has only two possible states, then it can be used in a quantum computer.

In fact, bearing this in mind, we have actually been using qubits throughout this book – without realising it. The example of the particle which can have only one of two colours – red or blue – forms a two-state system which represents a qubit.

One real-world example of a qubit arises from the spin of a particle. In the classical world, if we measure a spinning ball, we might expect to find its axis of rotation pointing in any random direction. However, if we measure the spin of an electron (by noting its deflection as it passes through a magnetic field), we find that only two distinct outcomes are possible: *spin-up*, or *spin-down*.

Before it is observed, the particle can be in a superposition state of *spin-up* **and** *spin-down*, but when it is observed it will be found to be in only one of those states. This is the characteristic behaviour of a qubit, so the spin of a particle can be used to represent a qubit.

However, this is not the technology which is being used in modern quantum computers. A recent technology called the *superconducting charge qubit* is being used by IBM, Google, and Intel. In a charge qubit, the state of the qubit ($|0\rangle$ or $|1\rangle$) is determined by the presence or absence of individual electrons in extremely small pieces of superconducting material. In other words, the state of a charge qubit is determined by the presence or absence of an amount of electric charge. This technology allows several microscopic qubits to be included on the surface of a single integrated circuit (chip).

Superconductivity happens when the temperature of certain materials is reduced to close to absolute zero. As a result, the materials acquire zero electrical resistance, so electrons can pass without being hindered – creating an electric current which can continue forever. Superconductivity holds great appeal for quantum computing because an electron which is not interacting with its environment is an electron which can remain in a coherent state – avoiding decoherence. And decoherence is, of course, the enemy of a quantum computer.

In his book *Where Does The Weirdness Go?*, David Lindley considers this ability of superconductors:

> *A superconductor can behave like a large-scale quantum object, but most macroscopic systems – pieces of ordinary copper wire, pointers on detectors, cats – do not. The distinguishing feature of a superconductor is the orderliness or coherence of the motion of its electrons, which stands in contrast to the disorderly motion of electrons in a copper wire and, in the same vein, the random jiggling around of atoms in most large objects.*

A charge qubit is created by placing two microscopic electrodes – extremely fine pieces of superconducting material – onto the substrate of the chip. The electrodes are separated by an insulating barrier (a *Josephson junction*) which normally prevents electrons crossing between the electrodes. However, it is possible for an electron to transfer from one of the electrodes to the other by a process called *quantum tunnelling*. Quantum tunnelling occurs when the wavefunction of an electron (representing the probability of finding that electron) extends to the other electrode, allowing the electron to overcome the insulating barrier.

So there is a finite probability that an electron can tunnel from one electrode to the other, representing a change in state of the qubit. And the probability of that transition is determined by the voltage which is applied to the qubit.

On the following diagram of an actual IBM quantum processor, you can see that the five charge qubits (square shapes) are labelled from Q_0 to Q_4. The chip shown in this image is actually about five millimetres square:

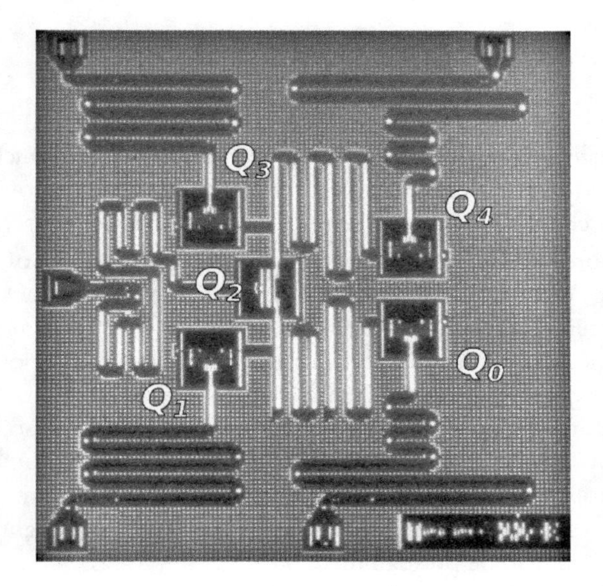

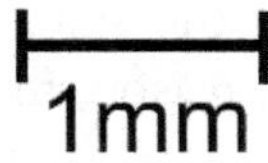

But there is something else amazing about this image …

… this is a photograph of one of the actual IBM quantum processors you will be programming later in this book!

The Bloch sphere

As explained in the previous section, the voltage which is applied to a charge qubit can control the probability of the qubit changing its state from $|0\rangle$ to $|1\rangle$, and vice versa. To be more precise, the voltage modifies the Hamiltonian of the qubit. We encountered the Hamiltonian operator back in Chapter Three during the discussion of the Schrödinger equation. In the simple form of the Schrödinger equation, it was explained how the Hamiltonian determines how the quantum state vector of a system varies with time. So, by modifying the Hamiltonian of a qubit, its state vector can controlled and rotated. That is how the state of an individual qubit can be modified, and therefore that is how a quantum computer can be programmed.

However, the state vector of a qubit rotates extremely fast. For this reason, the pulses of voltage which are sent to the qubit must be of extremely short duration, just a few hundred picoseconds (trillionths of a second). This places the signal in the GHz microwave frequency range.

As the IBM Q Experience FAQ explains: "Quantum gates are performed by sending electromagnetic impulses at microwave frequencies to the qubits through coaxial cables. These electromagnetic pulses have a particular duration, frequency, and phase that determine the angle of rotation of the qubit state around a particular axis of the Bloch sphere."

But what is the "Bloch sphere" mentioned in that quote?

To understand the Bloch sphere representation, first let us again consider the representation of a qubit in Hilbert space, with the state vector representing some mix of the $|0\rangle$ and $|1\rangle$ states:

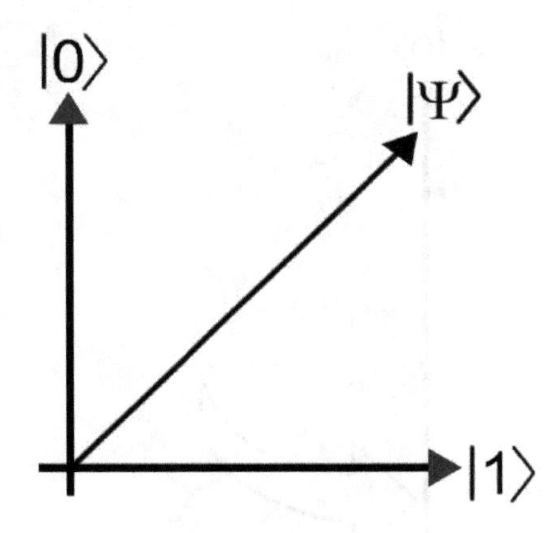

That is a nice, simple diagram to understand. However, just a little thought reveals that it cannot be a correct representation of the full picture. That is because – as was described in Chapter Five – coordinates in Hilbert space can be complex numbers. Therefore, any point in Hilbert space will have an associated phase angle. In particular, any quantum state vector will have a phase angle associated with it. However, there is no way for us to show the phase angle of the state vector on this diagram. The problem is that it is only a two-dimensional diagram, and to show an additional variable – the phase angle – associated with the vector would need us to add an extra dimension, and move to a three-dimensional diagram.

So that is what we have to do.

In order to make the move to a three-dimensional diagram, refer back to the previous diagram and imagine the horizontal $|1\rangle$ vector being pulled downwards so that it ends up pointing downwards as shown in the following diagram:

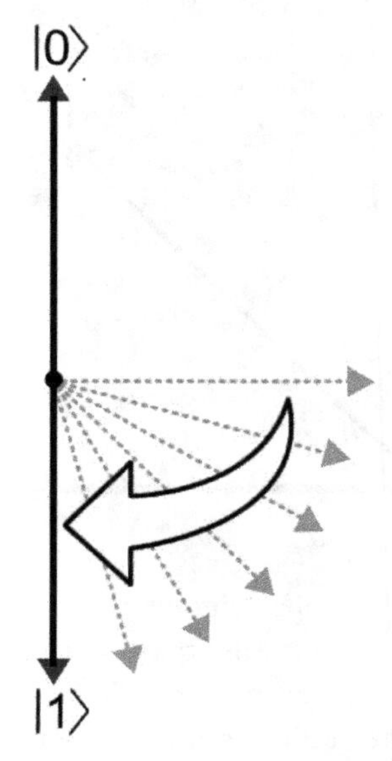

That now allows us to imagine the $|0\rangle$ vector as pointing to the top point on a sphere, and the $|1\rangle$ vector as pointing to the bottom point on a sphere, as shown in the following diagram:

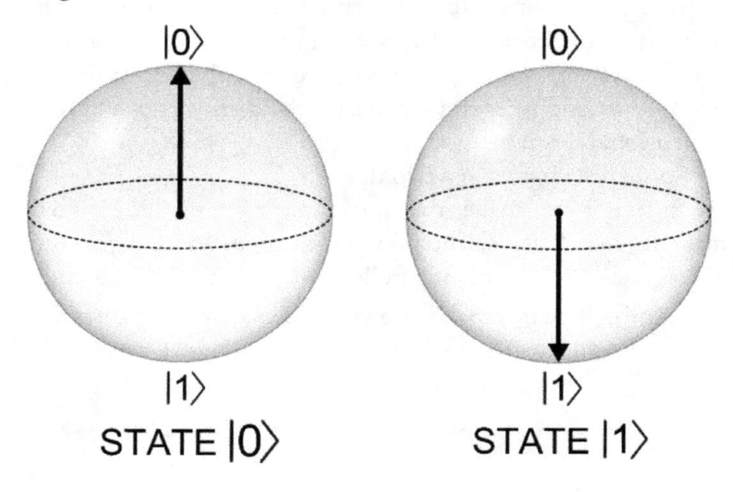

As you can see in the previous diagram, with the state vector pointing to the top of the sphere, that represents state $|0\rangle$, and with the state vector pointing to the bottom of the sphere, that represents state $|1\rangle$.

This method of representing a qubit is called the *Bloch sphere.*

For a qubit in a superposition state – an equal proportion of $|0\rangle$ and $|1\rangle$ – the state vector will point to the equator line, halfway between the $|0\rangle$ and $|1\rangle$ states. The following diagram shows a qubit in a superposition state:

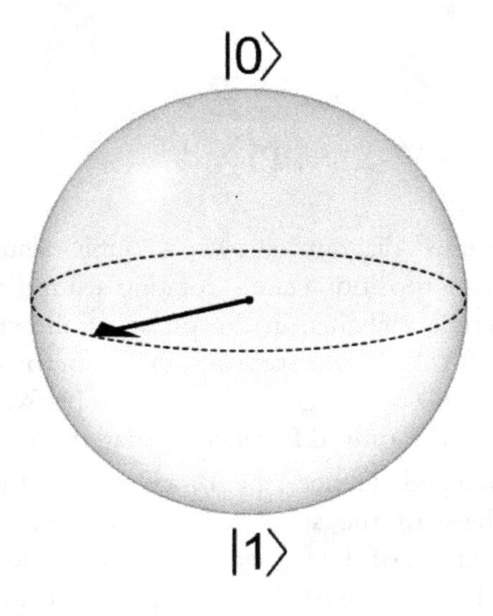

The big advantage of the Bloch sphere method comes when we consider the phase of the state vector. Because we are now working in three dimensions, the phase of the state vector can be represented by the rotation of the state vector around the vertical axis, as shown in the following diagram:

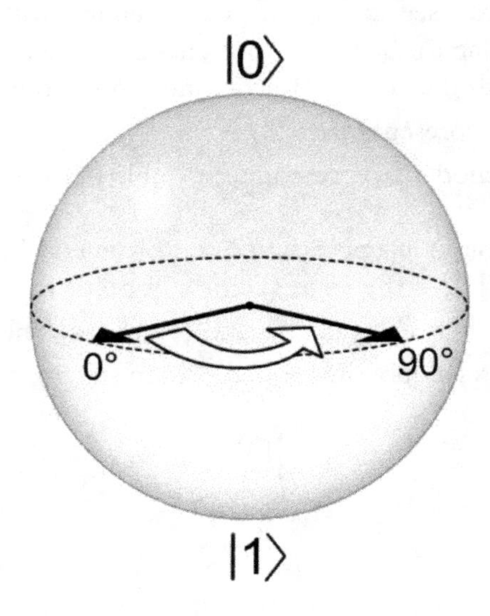

The previous diagram shows a qubit state vector – initially in a superposition state – rotating around the vertical axis to represent a change in the phase of the state vector from 0° to 90°. The qubit remains in a superposition state (the vector is still pointing to the equator, halfway between $|0\rangle$ and $|1\rangle$), the only difference is that the phase of the qubit has changed. Hence, the Bloch sphere method can show the phase of the state vector at any time: it is the amount of rotation of the vector around the vertical axis.

The Bloch sphere representation plays a central role in quantum computing. As we shall see in the next chapter, when you start programming a quantum computer you will be thinking in terms of rotating the state vector of a qubit around in the Bloch sphere.

Entanglement

Now we have come to the end of this chapter on the qubit, it seems like a good time to introduce the important concept of *quantum entanglement.*

In order to understand entanglement, we shall consider a qubit which is formed from the property of particle spin. Particles can have the property of spin, which has the units of angular momentum just like classical spin – the spin of a ball, for example.[10] However, when we measure the quantum spin of a particle we find it can only have two values: either *spin-up,* or *spin-down.* As these are the only two states we will find after measurement, this makes particle spin ideal to be used as a qubit. As an example, we might interpret the $|0\rangle$ state of our qubit as being spin-up, and the $|1\rangle$ state as being spin-down.

If a pair of photons is released by a common source, then the properties of those two photons will be related. This is because of the conservation laws. For example, the law of conservation of momentum states that the momentum **before** the photons were emitted must be equal to the total momentum **after** the photons were emitted. As the momentum before the photons was zero (nothing was moving), the total momentum after the photons are emitted must also be zero. This requirement can be satisfied if the photons are emitted in precisely opposite directions, because

---

[10] The connection between classical spin and quantum spin was clearly revealed in an experiment by Richard Beth in 1936 in which polarised light was shone onto a pendulum – causing the pendulum to rotate.

momentum is a vector quantity (like an arrow) and if we add two arrows pointing in precisely opposite directions then we get zero.

The law of conservation of angular momentum imposes a similar constraint of the spin of the photons. As the angular momentum before the photons were emitted was zero (nothing was spinning), then the angular momentum after the spin of the two photons is measured must also be zero. This requirement can be satisfied if the spin of the two particles is opposite: one of the particles being spin-up, and the other particle being spin-down.

When the properties of two separate particles are correlated in this manner, the particles are said to be *entangled.*

When the two particles are emitted, however, they are both in a superposition state, an equal mix of state $|0\rangle$ (spin-up) and state $|1\rangle$ (spin-down). Let us imagine what happens when we measure the spin of one of the particles and find it to be in the spin-down state. That implies that the state vector of that particle has rotated from its superposition state (an equal mix of states) to point toward the spin-down state. At that point, we then know with 100% certainty that when the spin of the second particle is measured it will be found to be spin-up (it must have the opposite spin value). This implies that the state vector of the second particle has also rotated – to point to the spin-up state. When the spin of the second particle is then measured, it will be found to be spin-up with 100% certainty.

These two rotations of the two state vectors (from the superposition state to spin-down and spin-up) are shown in the following diagram:

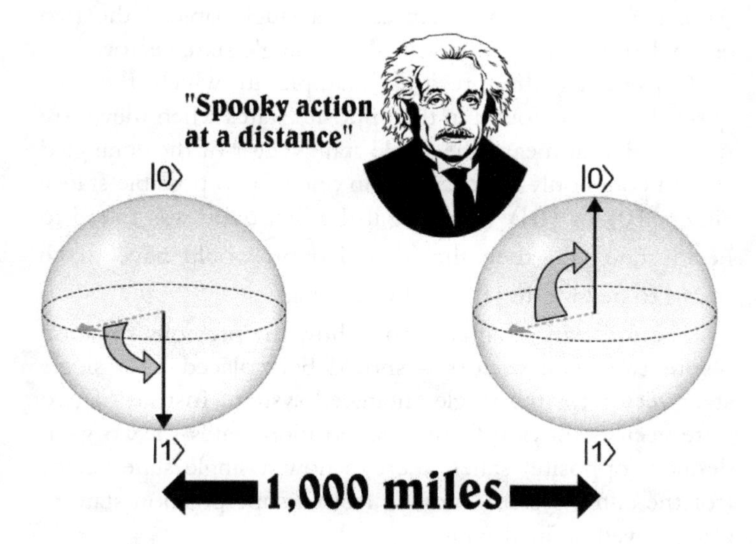

However, when Einstein heard of this behaviour predicted by quantum mechanics, he was horrified. This was because the two particles could, theoretically, be many thousands of miles apart when they are measured. Yet the rotation of the state vector of the second particle appears to happen instantaneously when the first particle is measured. Einstein's own theory of special relativity absolutely forbids this notion of an instantaneous effect being caused by a distant action. Special relativity states that nothing can travel faster than light, so there can be no instantaneous effects at distance. Einstein did not believe this prediction of quantum mechanics, and called it "spooky action at a distance".

The resolution of this mystery is quite simple – but it is fairly mind-bending.

The secret is that an entangled pair of particles can no longer be thought of as two independent systems – they must be considered as being a single system, a single object. As the Wikipedia page on quantum entanglement states: "A quantum state must be described for the system as a whole."

And, because they now represent a single object, the two entangled particles are described by a single state vector.

Considering the previous example in which the two qubits had to be found in the opposite states when they were measured, that means the single state vector of the entangled system could only be measured in one of two possible states: either $|10\rangle$ or $|01\rangle$ (meaning if the first qubit was found to be in state $|1\rangle$, then the second qubit would have to be found to be in state $|0\rangle$, and vice versa).

The following diagram shows how the previous situation – with two state vectors – should be replaced by a single state vector for the single entangled system. Instead of two state vectors moving from superposition states to two well-defined (opposite) states, there is now a single state vector (for the entire system) moving from a superposition state to a single well-defined state:

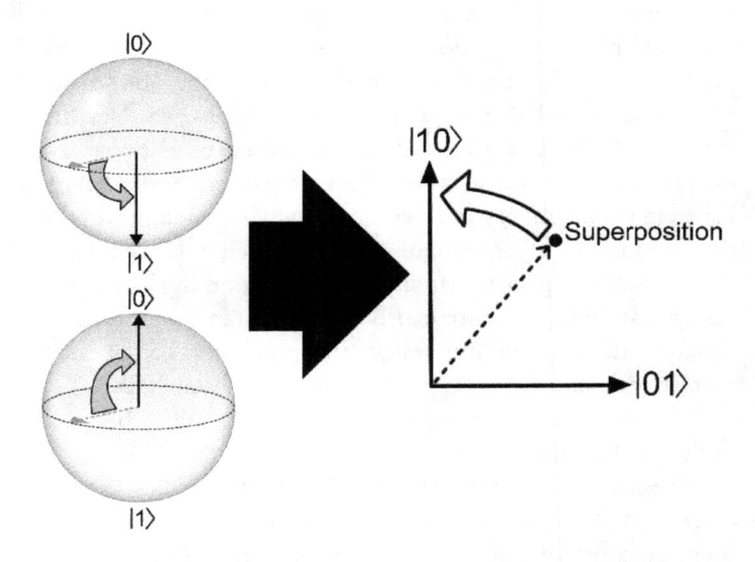

And this model means that all of the old problems about faster-than light causality and “spooky action at a distance”

disappear. We are now dealing with a single object, a single state vector. There is no longer a question of "which state vector causes the other state vector to move" because there is only one state vector. If you like, there is no longer a question of "spooky action at a distance" because if there is only one object then **there is effectively no distance!**

However, this neat resolution does imply that entangled particles separated by a great distance must be considered as being a single object. Admittedly, then, if there was one aspect of quantum mechanics that might be considered "weird", then it surely would be entanglement.

Later on in this book, we will be seeing that quantum entanglement plays a vital role in the functioning of quantum algorithms (programs) on a quantum computer.

This book has now covered the fundamentals of the theory of quantum mechanics. So all we need now is a real quantum computer …

8

THE IBM Q EXPERIENCE

The main headquarters of IBM Research is located in the small wooded town of Yorktown Heights, about thirty miles north of New York City. The center is named after the founder of IBM, Thomas J. Watson.

Some of the breakthrough technologies invented at the center include dynamic random access memory (or DRAM – described in my previous book), and the FORTRAN programming language. Benoit Mandelbrot also discovered the famous Mandelbrot set while working for IBM at Yorktown Heights. IBM may have a reputation for grey corporatism, but that is far from the truth. The most innovative companies are judged on how many patents they are awarded. IBM has far more patents for its inventions than any other American company. In contrast, Apple do not make the top ten. You see, there's science – and there's fashion. Don't get them mixed up.

It is in the Yorktown Heights facility where IBM's quantum computer research is located. In a remarkable feat of generosity, IBM have launched the IBM Q Experience which makes their quantum computers available for use by businesses, universities, and even the general public. The quantum computers can be accessed over the internet ("cloud-based") from their website. The IBM Q Experience is free for all users, which means you can get an account at their website and start using a real quantum computer today.

According to Andrew Houck, Professor of Electrical Engineering at Princeton University: "Thanks to this incredible resource that IBM offers, I have students run actual quantum algorithms on a real quantum computer as part of their assignments! This drives home the point that this is a real technology, not just a pipe dream. What once seemed like an impossible future is now something they can use from their dorm rooms."

You can easily find the IBM Q Experience website by Googling it, or going to the site directly:

http://quantumexperience.ng.bluemix.net

You will need an account before you can start experimenting with the quantum computer. On the top menu bar, you can Sign In if you already have an account, or Sign Up if you do not have an account.

When you have your account, Sign In, which will then take you to the homepage of the IBM Q Experience. There are useful resources on this page, including an excellent Beginner's Guide, and the Full User Guide.

To start experimenting on the quantum computer, click on "Composer" on the top menu bar, which takes you to the Composer window:

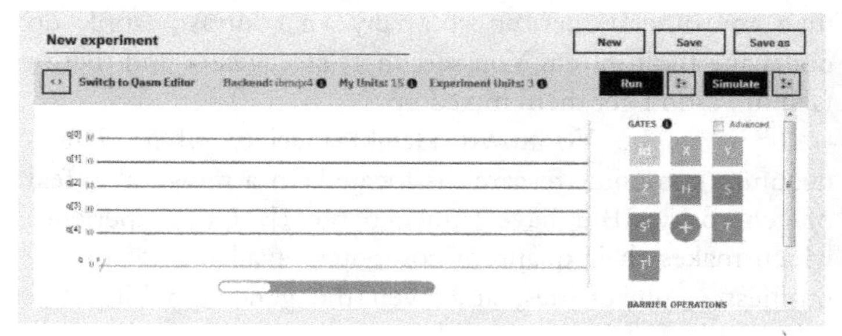

In this Composer window, you can see the five horizontal lines, reminiscent of a piece of sheet music (hence the name "Composer").

Each of the five horizontal lines represents one qubit (because the quantum processor available to you currently has five qubits – see the photograph of the actual five-qubit IBM quantum processor in Chapter Seven). You can see on the left side of each of the lines that each qubit starts in state $|0\rangle$.

Individual processing units which modify the state of each qubit are called *quantum gates*. You can see the gates listed as coloured square boxes on the right-hand side of the Composer:

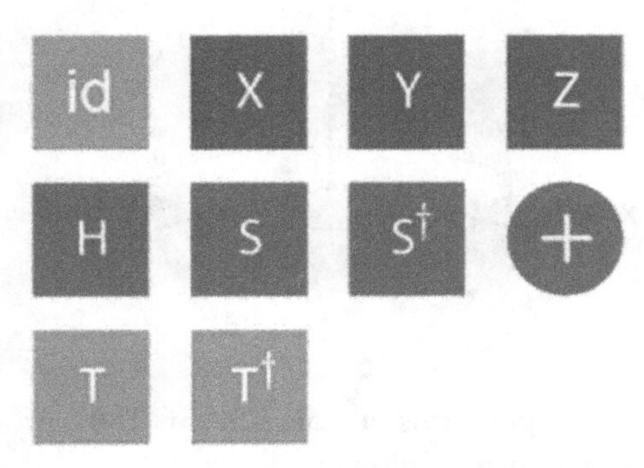

As the qubits pass along the horizontal lines from left to right, the qubits are processed by these gates. The final structure then forms what is known as a *quantum circuit*. The combination of gates provided for you by IBM form the standard set of gates with the whole package being known as the *standard circuit model*. Because of this, even if IBM one day discontinue the Q Experience, your experience of using it – and the standard circuit model described in this book – should remain relevant.

Let us now consider a few of these quantum gates, and show how they can be used to modify the state of a single qubit.

X: The bit-flip gate

The first quantum gate we will be considering is the X gate. The X gate is listed as one of the square boxes in the IBM Q Experience (see the previous diagram).

As stated earlier in this book, these quantum gates work by rotating the quantum state vector of a qubit. These rotations can be understood by considering the Bloch sphere as having three axes, as shown in the following diagram:

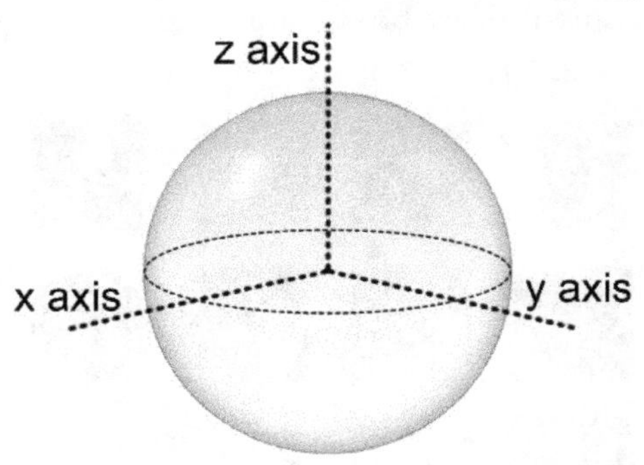

The X gate performs a rotation of 180° around the X-axis, as shown in the following diagram:

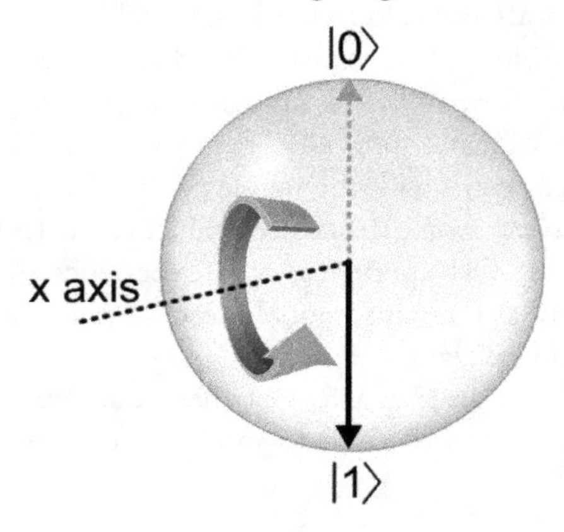

As you can see from the previous diagram, the effect of this rotation around the X-axis is to invert (or "flip") the state vector from $|0\rangle$ to $|1\rangle$ (or vice versa from $|1\rangle$ to $|0\rangle$).

The use of the X gate is shown in the following video which I recorded using the IBM Q Experience. You might want to sign-in to the Q Experience and perform the same experiment, or else you can just watch the video.

Here is a link to the video on YouTube:

http://tinyurl.com/quantumxgate

In the video, you will see that I actually use the simulator rather than the real quantum computer. You might find it more convenient to use the excellent simulator for development purposes. This is because a real quantum processor has qubits on a two-dimensional plane – the surface of the chip – so not every qubit can be connected to every other qubit (refer back to the photograph of the actual IBM processor in Chapter Seven). This places some limitations on which qubit can be connected to which other qubit, and you might find this rather frustrating when you are developing on the real quantum computer (it will sometimes not give you the option of connecting certain qubits to certain other qubits). In contrast, if you use the simulator, there are no restrictions on connectivity. Also, if you use the simulator, you get your results immediately, rather than having to wait for your experiment to be queued before it is run on the real quantum computer.

However, for our next experiment, we will be using the real quantum computer …

H: The Hadamard gate

The Hadamard gate is one of the most useful quantum gates. It is the gate which is used for placing a qubit into a superposition state, an equal mix of state $|0\rangle$ and $|1\rangle$.

You can see the effect of the Hadamard gate in the following diagram of the Bloch sphere. The effect of the Hadamard gate is to rotate the state vector around the diagonal dashed line:

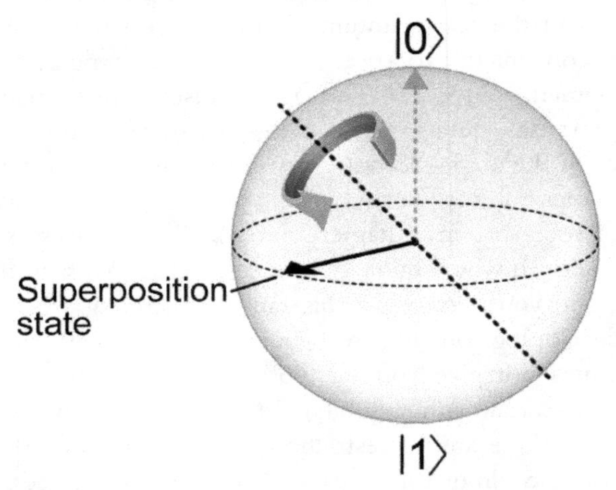

As you can see from the previous diagram, the effect of that rotation is to rotate a vector which is in state $|0\rangle$ (pointing to the North Pole) so that it points to the equator, a superposition state halfway between state $|0\rangle$ and $|1\rangle$.

The use of the Hadamard gate is shown in the following video which I recorded using the IBM Q Experience, this time running on a real quantum computer.

Here is a link to the video on YouTube:

http://tinyurl.com/gatevideo

Gates which modify phase

All of these quantum gates represent rotations of the state vector in the Bloch sphere. The last set of these gates represent rotations around the vertical Z-axis. As was explained in the previous chapter, rotations around the Z-axis represent a change in the phase of the qubit:

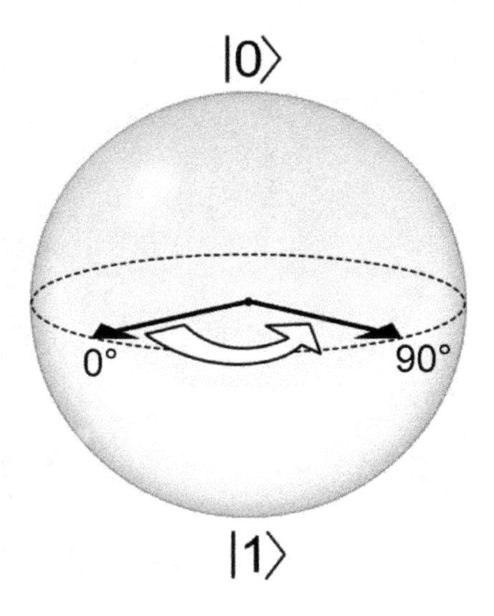

The Z gate performs a rotation of 180° around the Z-axis (just as the X gate performs a rotation of 180° around the X-axis, and the Y gate performs a rotation of 180° around the Y-axis).

Additional quantum gates in the IBM Q Experience include the S gate (which performs a rotation of 90° around the Z-axis – as shown in the previous diagram), and the T gate (which performs a rotation of 45° around the Z-axis).

So all three of these gates – Z, S, and T – act to change the phase of the qubit.

Gates which modify phases

[illegible] these quantum gates represent rotations of the state vector in the Bloch sphere. The last one, the X gate, [illegible] rotations around the x-axis [illegible] was explained in the previous chapter; [illegible] rotations around the Z-axis represent a change in the phase of the qubit.

The [illegible] [illegible] [illegible] [illegible] the [illegible] rotation [illegible] the Z-axis [illegible].

Add [illegible] [illegible] around the [illegible] [illegible] [illegible] the [illegible] [illegible] [illegible] [illegible] rotation [illegible].

So all these of these [illegible] Z [illegible] the [illegible] and [illegible].

9

MULTIPLE-QUBIT QUANTUM GATES

In the previous chapter we considered the simple quantum gates which take a single qubit as input, and output a single qubit. However, in this chapter, we will see that if we want to do anything interesting with our quantum computer we will require more complex gates which can take two qubits as input, allowing us to compare and combine the values of those qubits.

These type of multiple-input gates are the simple logical units which form the basis of today's conventional classical computers. So let us start this chapter by examining those conventional logic gates.

Boolean logic

Since the time of Aristotle in 350 B.C., logical reasoning had been the preserve of the philosophers, with their reasoned arguments. For example:

Socrates is a man.
All men are mortal.
Therefore, Socrates is mortal.

In the case of this argument, logical reasoning is used to prove that the statement "Socrates is mortal" is a true statement.

In 1847, the British mathematician George Boole wrote a book entitled *Mathematical Analysis of Logic* which attempted to place logic on a firm mathematical footing. George Boole realised that he could create a mathematical system of logic by substituting the number 1 for TRUE and the number 0 for FALSE. These TRUE and FALSE numerical values could then be combined according to strict mathematical laws. The ambiguous arguments of the philosophers could then be replaced by clear, unarguable, mathematical operations.

George Boole's method is now called *Boolean logic.*

But what George Boole could never have anticipated was the role that his Boolean logic would play in digital computers (which, of course, were not around in the mid-19th century). At their core, digital computers are based on binary arithmetic, with data bits which can only take the values of 0 or 1. These values, therefore, were a perfect match for the TRUE and FALSE values of George Boole's logical system. When digital computers were invented, George Boole's system was seamlessly integrated. As a result, Boolean logic and Boolean logic gates now form the basis of today's digital computers. The logical thought processes of philosophers such as Aristotle could now be replicated by an automated process – with Boolean logic essentially forming the "thought processes" of the modern computer.

So let us now consider the range of available Boolean logic gates. A logic gate is a device which takes one or more inputs (which can be either 0 or 1), performs some logical operation on those inputs, and produces a single output (which will be either 0 or 1).

Firstly, there is the NOT gate. The NOT gate is unique among Boolean gates in that it only takes one input (which would be either 0 or 1). The output of the gate is then the inverse of the input (the output is 1 if the input is 0, and vice versa).

Here is the symbol for the NOT gate:

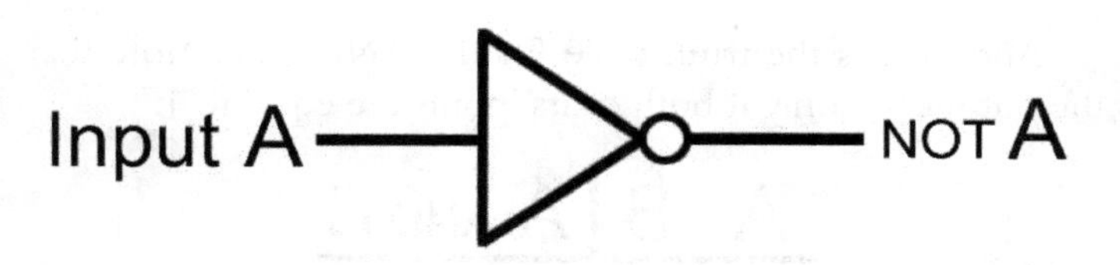

The behaviour of a Boolean gate can be described by a *truth table*, as shown in the following truth table for the NOT gate:

A	NOT A
0	1
1	0

You can see a truth table is divided into a left-hand side and a right-hand side. On the left-hand side, all the possible combinations of inputs are listed, and the right-hand side lists the output which will be produced by the gate when it receives those inputs. For the NOT gate, you can see that the output is simply the inverse of the input (the output is 1 if the input is 0, and vice versa).

All the other Boolean gates take two or more inputs. Let us consider the AND gate next.

The AND gate gives an output of 1 only if **both** of its inputs are equal to 1. Here is the symbol for the AND gate:

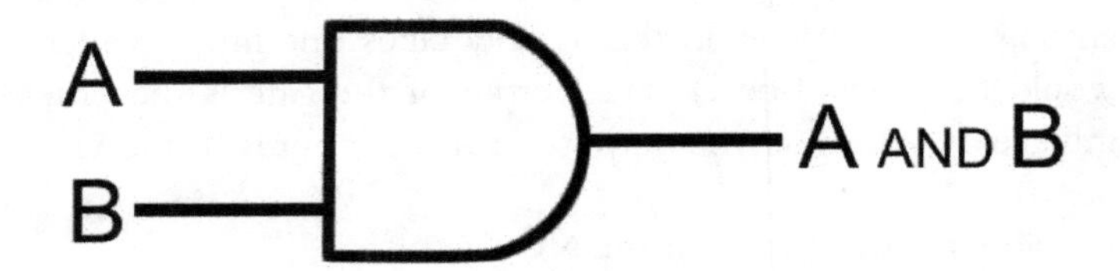

And here is the truth table for the AND gate. Note that the output is 1 only if both of its inputs are equal to 1:

A	B	A AND B
0	0	0
0	1	0
1	0	0
1	1	1

Now let us consider the OR gate. The OR gate gives an output of 1 if **either** input A **or** input B is equal to 1. Otherwise, the output of the OR gate is 0.

Here is the symbol for the OR gate:

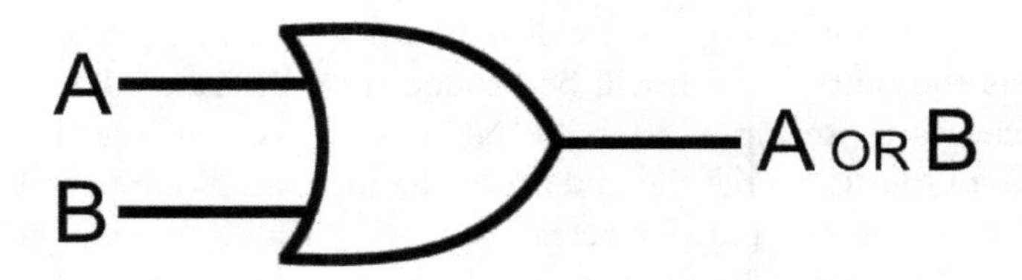

And here is the truth table for the OR gate. Note that the output is 1 only if at least one of the inputs is equal to 1:

A	B	A OR B
0	0	0
0	1	1
1	0	1
1	1	1

Next, there are the NAND and NOR gates. The NAND gate is similar to the AND gate except its output is inverted (the opposite of the AND gate output). While the NOR gate is similar to the OR gate except its output is inverted (the opposite of the OR gate output).

Here is the symbol for the NAND gate:

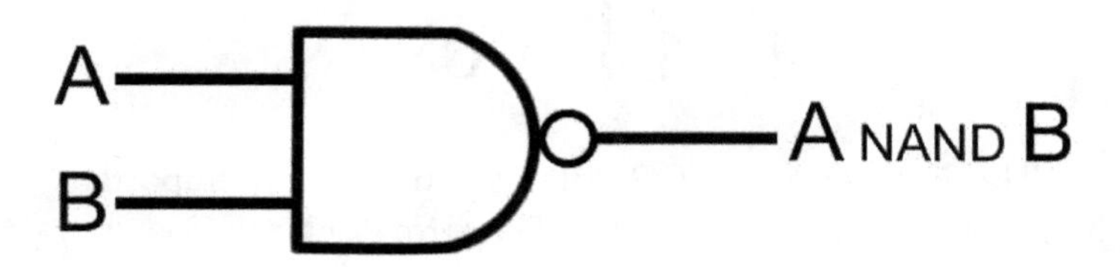

And here is the truth table for the NAND gate. Note that the output is the opposite of the output of the AND gate:

A	B	A NAND B
0	0	1
0	1	1
1	0	1
1	1	0

Finally, here is the symbol for the NOR gate:

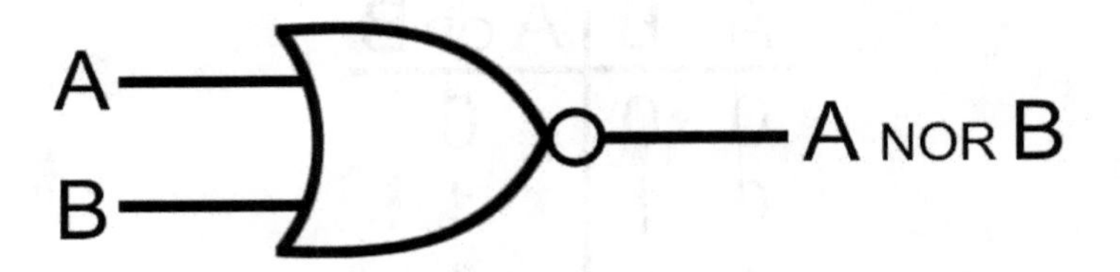

And here is the truth table for the NOR gate. Note that the output is the opposite of the output of the OR gate:

A	B	A NOR B
0	0	1
0	1	0
1	0	0
1	1	0

The NAND and NOR gates are particularly important because they are *universal* gates. The term "universal" is used when an object can do anything, it can behave like any other object. A universal logic gate can behave like any other logic gate, and that makes it very useful and very powerful.

Let us consider examples of how a NAND gate or a NOR gate can act like a different gate. Firstly, if both of the inputs to a NAND or NOR gate are the same, then the gate will act like a NOT gate. You might like to check that this is the case by looking back to the truth table for the NAND or NOR gate and seeing that when both inputs are 1 then the output will be 0, and when both inputs are 0 then the output will be 1. Hence, the (inverting) behaviour is the same as a NOT gate. This is shown in the following diagram:

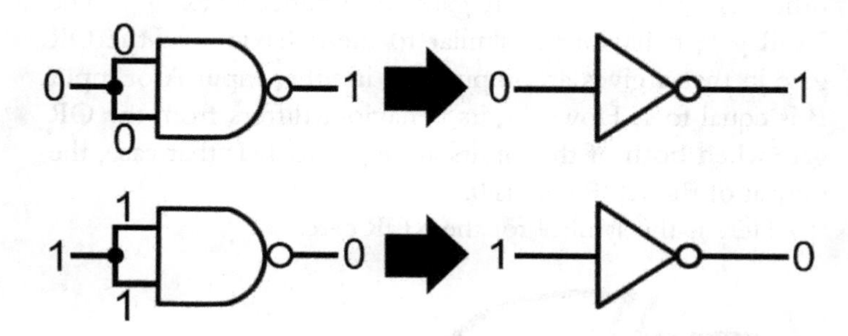

The previous diagram shows how a NAND gate with both its inputs tied together will behave like a NOT gate. This is one example of the universality of the NAND gate.

A more complex example of universality is presented in the following diagram. You might like to check from the truth tables that the following circuit of three NAND gates performs like an OR gate:

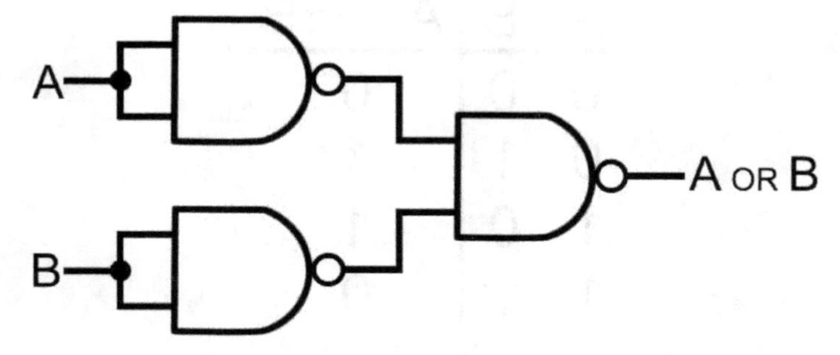

The previous diagram represents an example of how logic gates can be combined to create more complex circuits. We will be seeing that this is an essential skill for programming a quantum computer – using the quantum equivalent of Boolean logic gates. Fortunately, many IT professionals will already have experience of using Boolean logic gates, so that skill will be transferrable to quantum computing.

The final Boolean gate is the exclusive-OR gate, otherwise called the XOR gate (pronounced "ex-or"). The XOR gate behaviour is similar to the behaviour of the OR gate in that it gives an output of 1 if either input A or input B is equal to 1. However, its behaviour differs from the OR gate when **both** of the inputs are equal to 1. It that case, the output of the XOR gate is 0.

Here is the symbol for the XOR gate:

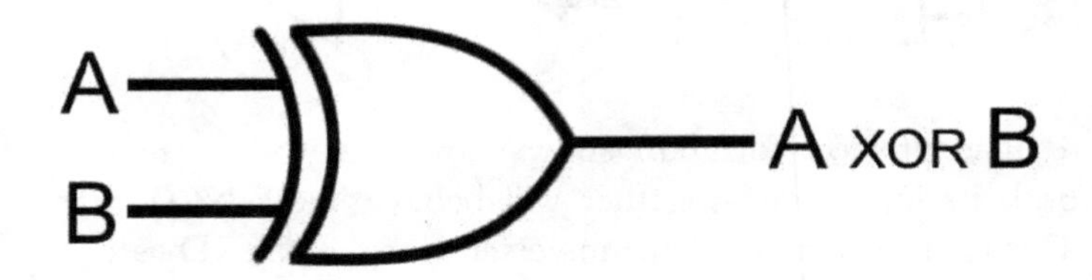

And here is the truth table for the XOR gate. Note that the output is the 0 when both of the inputs are equal to 1:

A	B	A XOR B
0	0	0
0	1	1
1	0	1
1	1	0

As we shall be seeing shortly, the XOR gate is of particular importance in quantum computing.

Unfortunately, the life of George Boole was to come to a rather tragic end. One rainy evening, Boole came in to his house with a very slight cold. His wife, unfortunately, had rather peculiar views about curing illnesses: she was convinced that the cure for an illness should resemble its cause. Because of this belief, she proceeded to continuously pour buckets of cold water over the poor man until he died from pneumonia at the age of just 49.

Reversible computing

Our next task is to implement these Boolean logic gates on a quantum computer. We would then have a guarantee that our quantum computer would be able to solve the problems which could be solved using a conventional classical computer. However, in this section we will see that implementing these Boolean gates on a quantum computer is rather tricky.

Let us first consider the AND gate. Here is the truth table for the AND gate which was presented earlier:

A	B	A AND B
0	0	0
0	1	0
1	0	0
1	1	1

As you can see from the previous table, the AND gate takes two inputs but has only one output. As a consequence, if you were only presented with the output of the gate, you would not be able to say what that the inputs had been with complete certainty. As an example, you can see from the previous table that if you knew the output of an AND gate was 0, you would not be able to tell if the inputs had been 0 and 1, or both inputs had been 0. In both of those cases, the output of the AND gate would have been 0. In other words, some knowledge – some *information* – would have been lost through the application of the gate.

This type of two-input Boolean gate is therefore not reversible. If all we know is the output, we cannot "wind the gate back" and get the input values.

This might not seem like a very big deal for a conventional computer. After all, logic gates in a conventional computer only ever have to work in the forward direction. However, it represents a serious problem for a quantum computer.

The problem is due to the nature of the Schrödinger equation which we considered in Chapter Three. If you remember, the Schrödinger equation determines the behaviour of the quantum state vector over time, and the Schrödinger equation states that this motion is proportional to the total energy of the system (the Hamiltonian). As a result, the rotation of the state vector is linear: smooth and predictable. This is called *unitary evolution.* This also means the motion of the state vector is completely reversible: you could theoretically "wind back" the rotation of the state vector to an earlier state. Because of this reversibility of the Schrödinger equation, all quantum logic gates must be *reversible.*

However, as we have just seen in the AND gate example, if information is lost then the gate is **not** reversible (the inputs cannot be regenerated from the outputs). Therefore, we can deduce that **no information can be lost in unitary evolution.**[11]

This means that, because the AND gate loses information and is therefore not reversible, it would appear it is not possible to implement an AND gate as a quantum gate.

The CNOT gate

The important point we have just discovered is that information is lost in an AND gate – and the other Boolean gates – and so it appears it cannot be implemented as a quantum gate. Does that mean we cannot make a quantum computer? Let us not give up yet.

Instead, we will now see that there is a clever trick we can use to make an irreversible logic gate reversible. The trick requires carrying-over one of the inputs to the outputs. If both of the outputs are then examined, it is possible to uniquely determine what the inputs were. The gate therefore becomes reversible, and it can be implemented in a quantum computer.

Let us first start with a gate which we have just determined is irreversible – the AND gate – and let us see if this trick works. The following diagram shows an AND gate with one of the inputs carried-over to become one of the outputs:

---

[11] This is the principle behind one of the greatest mysteries in physics: the black hole information loss paradox. See my second book for more details.

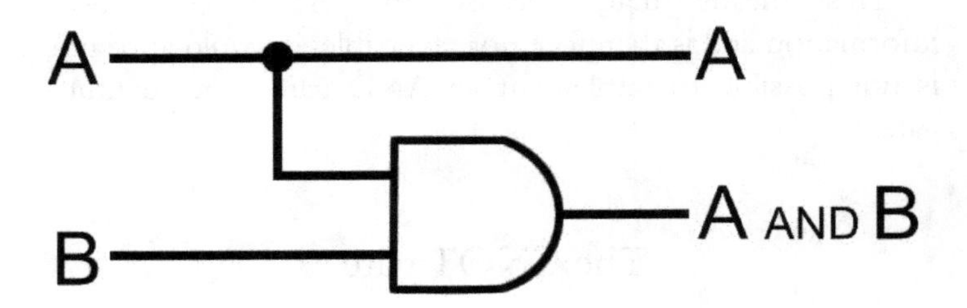

Let us examine the truth table for this circuit, now with the two outputs:

INPUTS		OUTPUTS		
A	B	A	A AND B	
0	0	0	0	Same outputs
0	1	0	0	
1	0	1	0	
1	1	1	1	

We can see from the previous truth table that it is still not possible to uniquely identify the inputs from the outputs. For example, you can see that if the outputs of the circuit are 0 and 0, you would not be able to tell if the inputs had been 0 and 0, or 0 and 1. So **information has been lost** and that means that this AND circuit cannot be implemented as a quantum gate.

This is not only the case for the AND gate – you might like to check that this problem also applies to the OR gate, the NOR gate, and the NAND gate.

But let us now consider the exclusive-OR (XOR) gate which was described earlier in this chapter. The following diagram shows an XOR gate with one of the inputs carried-over to the outputs:

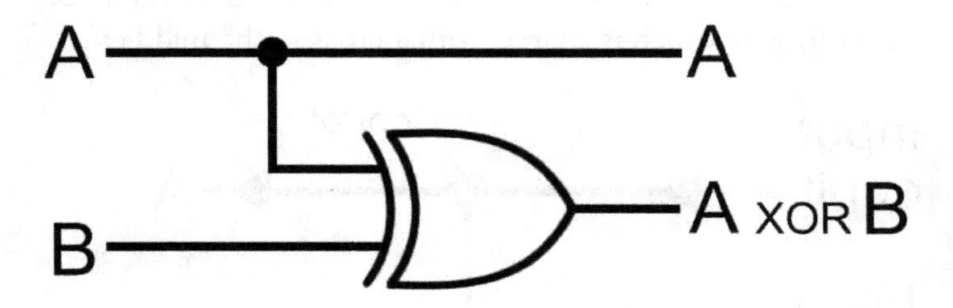

Let us examine the truth table for this circuit, now with the two outputs:

INPUTS		OUTPUTS	
A	B	A	A XOR B
0	0	0	0
0	1	0	1
1	0	1	1
1	1	1	0

Each of the four output lines are different

In this case, you can see that the two outputs uniquely identify the two inputs. For example, if the output pair is 1 and 1 then we know that the input pair was definitely 1 and 0. So no information is lost, which means this gate is reversible (the inputs can be regenerated from the outputs).

So the XOR gate with one of the inputs carried-over can be implemented as a quantum gate. That means we have managed to discover a reversible two-input quantum gate we can use in our quantum computer! Hurrah!

Here is a diagram of this XOR gate circuit as drawn on a quantum circuit diagram. You will see I have replaced the written "XOR" gate symbol with the standard symbol for the XOR gate which is a circle with a cross in the middle:

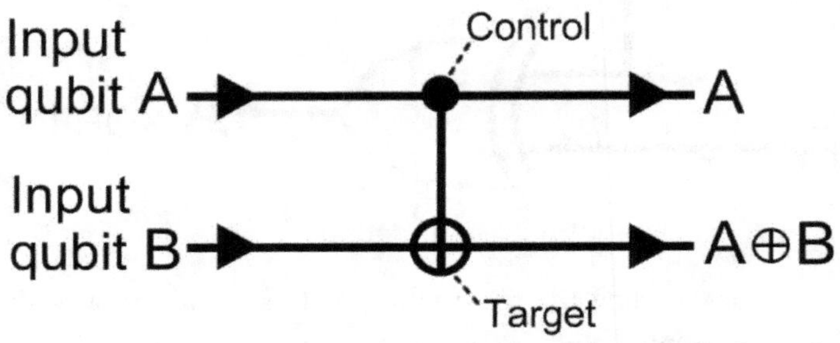

You will also see that qubit A has been labelled as the "control" qubit, and qubit B has been labelled the "target" qubit. The reason for this will now be explained, and we shall see it is because of the way the behaviour of this gate is interpreted in quantum computing.

This quantum version of the XOR gate is not called an XOR gate in quantum computing terminology. To see why that is the case, let us examine the truth table for the XOR gate in this circuit, and see if we can interpret it in a different way:

A	B	A⊕B
0	0	0
0	1	1
1	0	1
1	1	0

From the previous truth table, we can see that if qubit A is equal to 0, then the output is equal to the value of qubit B. But if qubit A is equal to 1, then the output is equal to the inverse of qubit B (the NOT operation applied to qubit B). In other words, the NOT operation is only applied if qubit A is equal to 1. Therefore, we could interpret the operation as a NOT operation which is controlled by qubit A.

And, indeed, this is how this XOR circuit is described in quantum computing: it is called the controlled-NOT gate, or CNOT gate. To recap, in a CNOT gate, the NOT operation is only applied on a qubit when the value of another qubit is equal to 1.

The qubit which controls the operation (qubit A in this case) is called the *control* qubit, and the qubit which is inverted (qubit B in this case) is called the *target* qubit (as shown on the earlier circuit diagram of the CNOT gate).

Here is a diagram of the CNOT gate as shown on a quantum circuit diagram:

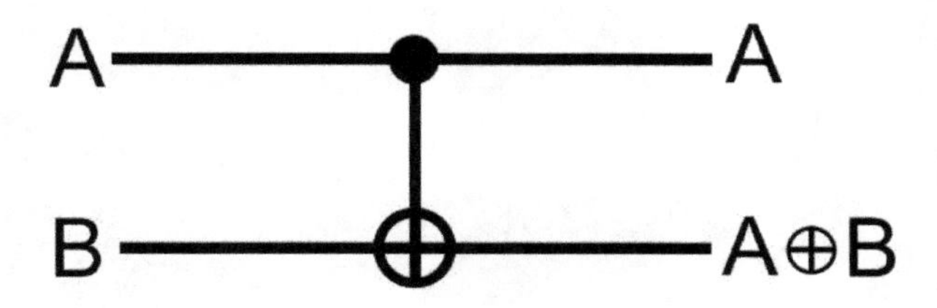

So we have discovered that the CNOT gate is the only reversible two-input Boolean logic gate which can be implemented as a quantum gate. This explains the layout of the available gates in the IBM Q Experience:

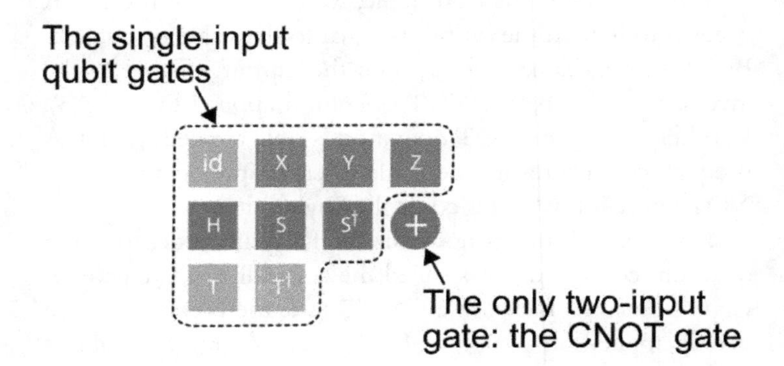

You can see in the previous image that the range of gates available to you in the IBM Q Experience (the standard circuit model) include the full range of single-input quantum gates, plus the only possible two-input quantum gate: the CNOT gate.

The Toffoli gate is your friend

However, this limited range of gates represents a problem. The problem is that we do not have any universal gates in our set of gates. As described in the earlier discussion on Boolean logic, a universal gate can be used to behave like any other gate. The CNOT gate is our only two-input quantum gate but it is not universal, which means it cannot be used to create all the other possible gates such as AND and NOR. If we want to be able to program anything on our quantum computer, we need to find some way of implementing those other Boolean gates.

A solution emerges when we consider expanding the CNOT gate to take an additional input. If you remember, the CNOT gate is a "controlled-NOT" gate, which means it is really a NOT gate which is controlled by another qubit. We can extend this picture by adding an extra level of "control" – we can control the controlled-NOT gate.

We can get a controlled-controlled-NOT (or CCNOT) gate by having an additional qubit control a controlled-NOT gate. Then, only if this additional qubit is in state $|1\rangle$ is the controlled-NOT operation performed.

Here is the symbol for the CCNOT gate, which is more commonly known as the *Toffoli gate*:

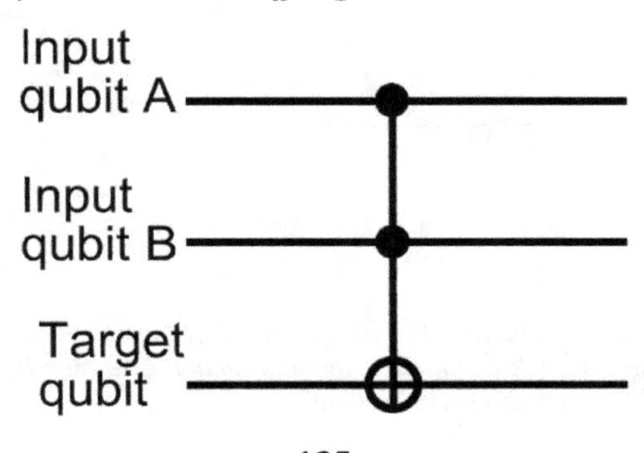

In other words, both input qubits must be in state $|1\rangle$ for the NOT operation to be performed on the third (target) qubit. If the third qubit is in initially in state $|0\rangle$ you can see that the Toffoli gate acts like an AND gate. That is because the third qubit will be inverted – changed to $|1\rangle$ – only if the other two qubits are both set to $|1\rangle$. And that is the characteristic behaviour of an AND gate (output equal to 1 when both inputs are 1).

What is more, it is easy to see that if the third qubit was initially in state $|1\rangle$ rather than state $|0\rangle$ then the Toffoli gate would act like a NAND gate rather than an AND gate. This is because if both input qubits were in state $|1\rangle$ then the target qubit would be flipped to be in state $|0\rangle$ – and that is the characteristic behaviour of a NAND gate (output equal to 0 when both inputs are 1).

That's great, because we know the NAND gate is universal, so we now have a universal set of quantum gates.

And what makes that such great news is that if we have a universal set of quantum gates then we have a guarantee that **we can calculate anything on our quantum computer which can be calculated on a classical computer!**

The Toffoli gate is not available as one of the gates within the IBM Composer. However, it is possible to program it using the available gates as shown in the following diagram:[12]

[12] This diagram can be found in Chapter Four of the quantum computation textbook *Quantum Computation and Quantum Information* by Michael Nielsen and Isaac Chuang.

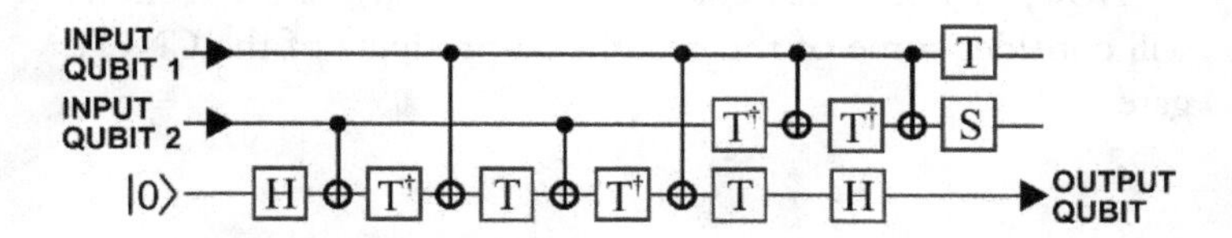

It is an interesting experiment to create the Toffoli gate as shown in the previous diagram inside the IBM Composer (I tried it, and you might like to try it too), but it clearly requires a lot of gates just to create the equivalent of a single Boolean AND gate. Fortunately, those thoughtful people at IBM have provided us with a simple form of the Toffoli gate for us to use. The developers at IBM have placed the previous arrangement of gates into a *subroutine* – a single gate which contains a series of many other gates – which means it is now a simple task to include the Toffoli gate on our circuit diagrams.

The use of the Toffoli gate subroutine is shown in the following video which I recorded using the IBM Q Experience. Here is a link to the video on YouTube:

http://tinyurl.com/gatevideo3

And here is a still from the video:

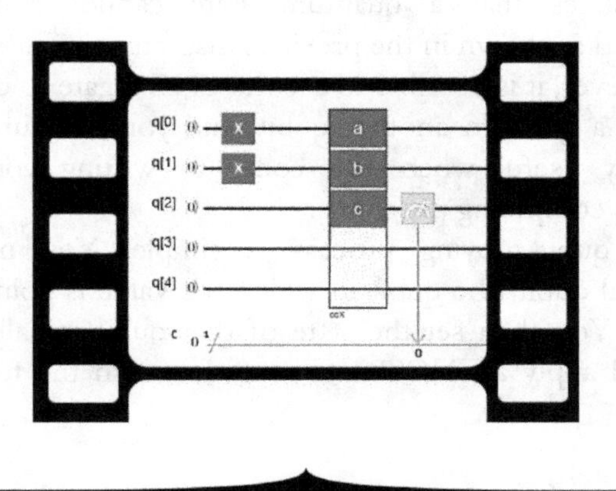

Now, to finish this chapter, the final two short sections will consider some of the useful side-products of the CNOT gate.

Copying a qubit

Sometimes while writing a quantum computer program you might find it useful to copy the value of a variable or qubit. You might imagine it would be easy to copy the value of a qubit in a quantum circuit diagram simply by using a T-junction:

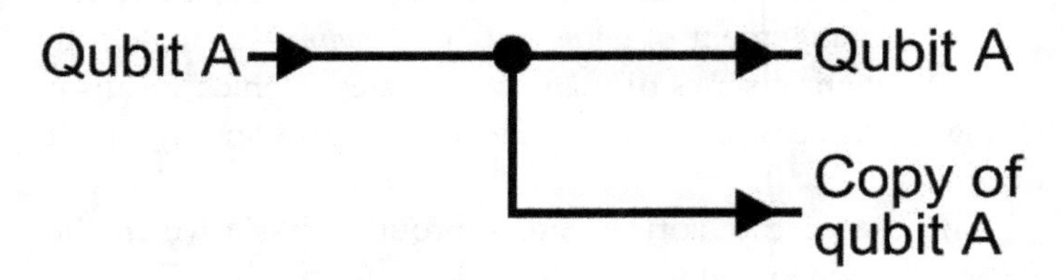

However, that is not the case: T-junctions such as the one shown in the previous diagram cannot exist on quantum circuit diagrams. This is due to the quantum *no-cloning theorem* which states that a quantum state cannot simply be duplicated as shown in the previous diagram.

However, it is possible to use the CNOT gate to copy the value of a qubit to another qubit, and you will surely find that very useful when you come to writing your own quantum computing programs.

The qubit-copying process is simple. You need an additional qubit (the qubit to which the value is going to be copied). You then set the state of that qubit initially to be $|0\rangle$, and apply a CNOT gate – as shown in the following diagram:

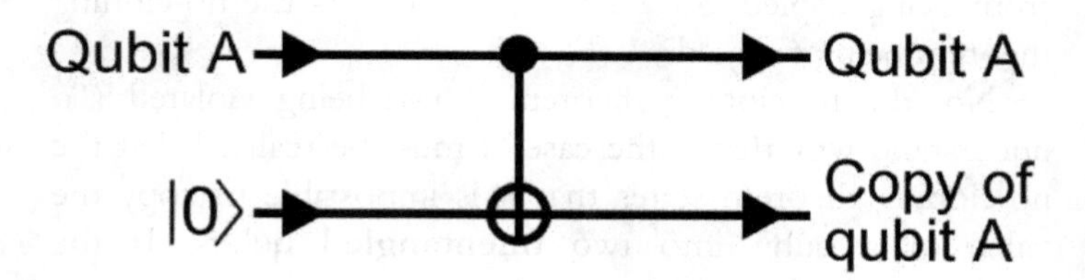

You can see from the diagram that if the input qubit A is equal to $|0\rangle$, then the NOT operation will not be applied (because that is how the CNOT gate works), and so the output copy on the second qubit will remain at its initial state of $|0\rangle$. But if the input qubit A is equal to $|1\rangle$ then the state of the second qubit will be inverted by the CNOT gate to become $|1\rangle$. So, in both cases, the value of the input qubit is copied to the output qubit.

Entanglement and the CNOT gate

But the most important side-product of the CNOT gate is that it entangles the two input qubits.

If you remember, the definition of a CNOT gate is that it inverts the value of a qubit (applies the NOT operation) only if the state of a particular different qubit is $|1\rangle$. It is clear, then, that the use of the CNOT gate results in a dependency of the value of one qubit with the value of another qubit. This dependency is the very definition of entanglement, so the CNOT gate inevitably entangles its two input qubits. In fact, the CNOT gate is the method by which qubits are entangled in a quantum computer.

And this raises an interesting point …

In the previous section we have just seen how the CNOT gate can be used to copy the value of a qubit, but it was also stated that the quantum no-cloning theorem prevents qubits from being copied. So what is going on? Is the no-cloning theorem being violated?

No, the no-cloning theorem is not being violated. To understand why that is the case, it must be realised that the no-cloning theorem states that it is impossible to copy the value of a qubit into two **unentangled** qubits. If the resultant qubits are entangled, then the no-cloning theorem does not apply. This is because, as was explained in Chapter Seven, when two objects are entangled then they represent a single object as far as quantum mechanics is concerned. So, as far as quantum mechanics is concerned, the two qubits have not been cloned at all – they are still one object!

10

QUANTUM ALGORITHMS

The experiments we have performed so far have not exploited the full power of the quantum computer. Indeed, in his *Lectures on Computation*, Richard Feynman covers the basics of quantum computing and reversible computing as we have covered so far in this book, but Feynman concludes:

> *What we have done is only to try to imitate as closely as possible the digital machine of conventional sequential architecture. What can be done, in these reversible quantum systems, to gain the speed available by concurrent operation has not been studied here.*

The phrase "concurrent operation" is a reference to parallel processing. So Feynman is stating that in order to extract the maximum processing power from our quantum computer, we need to exploit the potential of the quantum superposition state to perform calculations in parallel – just as Feynman himself exploited the power of parallel processing with the IBM mechanical calculating machines back in Los Alamos.

If you remember, at Los Alamos, Feynman could have multiple data in his line of mechanical calculating machines by colour-coding the punched cards so he could keep track of the cards which were at different stages of the process. The principle behind a quantum computer is similar: multiple copies of the data can be held in the central processor at the same time, and those multiple copies of the data can be processed simultaneously. This enables the potentially vast increase in processing speed associated with parallel processing.

Let us explore this process in more detail.

From bits to qubits

At the core of a conventional computer lies the central processing unit (CPU) which, in most computers, takes the form of a microprocessor (a single integrated circuit). Within the microprocessor are a series of registers. A register is a line of bits, usually representing a single binary number. It is common for a single register to have maybe 8 bits, or 32 bits, but it is 64-bit registers which are usually found in modern computers ("64-bit processing").

The following diagram shows the decimal number 153 stored in an 8-bit register in a microprocessor in binary form (10011001). You can see each of the bits labelled from d_0 to d_1:

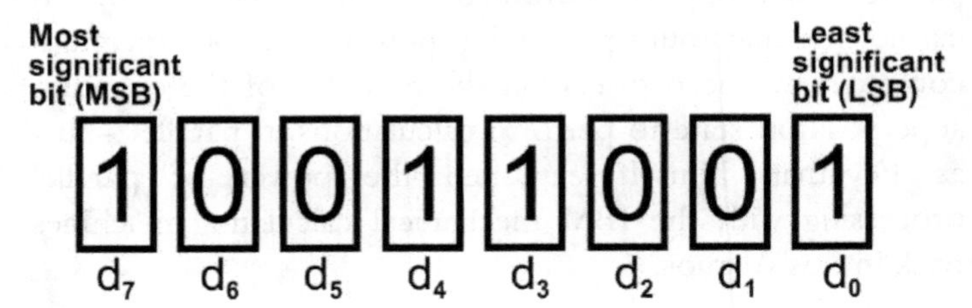

The contents of a register represents the piece of data which the computer is currently processing. Various operations can be applied to the data in the register as the microprocessor steps through a computer program line-by-line. These operations would generally be Boolean logic operations such as AND and NOR which we considered in the previous chapter. In this way, even complicated computer programs are translated into very many simple Boolean operations which are then applied to the data stored in a register. In a modern computer, these operations can be applied in nanoseconds. That is incredibly quick, **but we are still dealing with only one piece of data at a time.** The only reason our computers are as fast as they are is because of their incredibly fast clock speed (speed at which a single operation is performed). Otherwise, conventional computers have a processing bottleneck, only being able to perform one instruction at a time.

However, a quantum computer promises to eliminate that processing bottleneck by holding multiple copies of its data in a superposition state. The inputs of a quantum computer are not composed of classical bits – they are composed of qubits. As an example, you will remember that the IBM Q Experience quantum processor we have been using so far in this book has five input qubits. If you look carefully in the composer window you will see that these are labelled from q[0] to q[4], so we might interpret those qubits as making a five-qubit register.

And, of course, each of those qubits is capable of being in a superposition state which effectively means it is holding both the value zero and one **at the same time.** We now start to see how a quantum computer can overcome the processing bottleneck of a conventional computer.

To explain the advantage of a quantum computer, let us first consider an 8-bit classical register. With eight bits, the register is capable of holding 256 different binary numbers

($2^8 = 256$). However, the register can only hold one of those 256 numbers at a time. In contrast, each of the qubits in a quantum computer can be both zero and one at the same time, and so a quantum register can consider all 256 different numbers at the same time.

As in a classical register, we would then apply logic gates to the qubits is order to process their values (stepping through each step of a computer program). But, of course, in a quantum computer we would apply our quantum gates (such as the X, H, and CNOT gates described in the previous two chapters) instead of the Boolean logic gates (AND, OR, and NOT) we would apply to a classical register. And those quantum gates would be being applied to multiple copies of the input data at the same time – you get so much more processing power for your buck.

And, because the number of possible input combinations rises exponentially (according to 2^N, where N is the number or qubits), a quantum computer gets very powerful very quickly. As just described, with eight input qubits there would be 2^8 different input combinations, which is equal to 256. And the quantum computer would be able to consider all of those 256 different input combinations at the same time. With sixteen qubits there would be 65,536 different input combinations. With 24 qubits there would be over 16 million different input combinations. And with just 300 qubits, a quantum computer would have more input combinations than there are atoms in the universe.

Let us now consider the problems we could solve using that tremendous quantum computing parallelism.

The inverse problem

What do we mean by a "problem"?

When we think of a typical problem, we probably think of a question which needs an answer. As a common example, I might ask: "Where have I left my reading glasses?". A different example from mathematics might be: "What is the answer when we multiply 47 by 31?"

So these represent the sort of problems we might encounter in everyday life: problems when we know the question and need to find the answer. However, there is a completely different type of problem. These are the problems when we know the answer, and we need to go back to determine the question. It might sound easy, but it's not.

Determining the question when we know the answer is called the *inverse problem.*

In general, to solve the inverse problem, we need to determine the reverse of the procedure which we would use to calculate the answer from the question. In that way, we calculate the *inverse transformation.* The inverse transformation would be the reverse of the forward transformation, with the forward transformation being the normal process which is used to calculate the answer from the question.

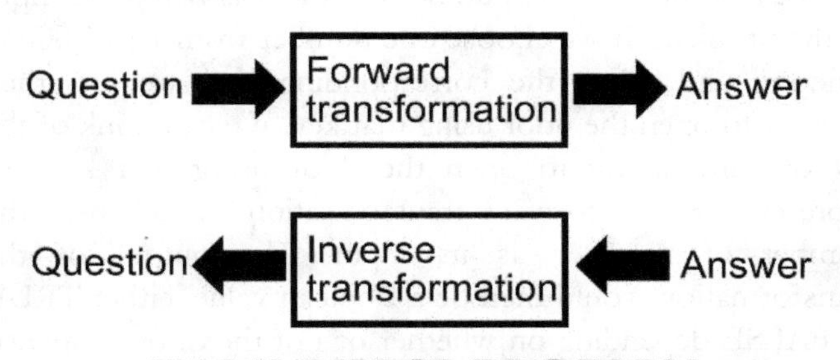

THE INVERSE PROBLEM

However, sometimes it might not be possible to calculate the inverse transformation, or it might be incredibly difficult. In other words, there is no way to directly link the answer with one particular question. Here is an example of such a problem.

You are faced with a locked door. On a table in front of you are a thousand different keys. Each key has a label attached to it with a number from 1 to a 1,000, clearly identifying each individual key.

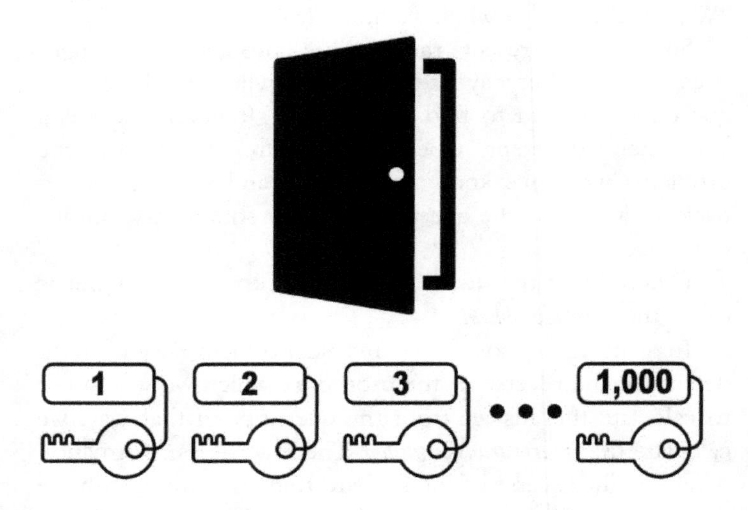

The "problem" in this case is how to open the door.

We can think of the number on a key as being the input to the problem. If we choose one number from 1 to 1,000 at random, and select the corresponding key, we can then attempt to open the door using that key. We can think of the act of "attempting to open the door using that key" as representing the "forward transformation" which uses the number of the key as its input. The output of the transformation would then be a Boolean value, either TRUE or FALSE depending on whether or not the door is opened by the key.

So the forward transformation in this case is straightforward. When you are asked "Does key number X open the door?", you select the key with that number (X), you insert the key into the lock on the door, and you see if the key opens the door. You get your answer to the question.

So we have a clear forward transformation working forwards from the question to the answer. But what is the reverse transformation, working backwards from the answer to the question? To be precise, if we are presented with the answer TRUE – that the door is open – is there any way we can work backwards from that answer (which is simply TRUE) to determine the number of the key which opened the door?

No, it is clear there is no inverse transformation. If we are presented with an open door, we cannot tell which key opened the door.

So, in the absence of an inverse transformation, do we give up, defeated?

No, of course we don't give up. We don't give up because we do at least have the forward transformation – we are able to test an individual key to see if it opens the door. We could then exhaustively test every one of those thousand keys to discover which key opens the door. Then, when we are presented with the inverse problem "Which key opened the door?" we will be able to provide an answer to that inverse problem.

That sounds straightforward: it appears that as long as we are in possession of the forward transformation we will be able to solve any inverse problem – it will just take a long time to test all the possible inputs. But what happens if there are more than a thousand keys? What happens if there are a million keys? What happens if there are a trillion keys? What happens if there are as many keys as there are atoms in the universe?

In those situations – with so many possible inputs – it appears we cannot use this method of exhaustively testing all possible inputs. **However, this method of considering all possible combinations of inputs is precisely the sort of thing that quantum computers do so well!** As we have seen, if the input qubits of a quantum computer are prepared in a superposition state, then **a quantum computer can check all possible combinations of inputs at the same time.** And, as was explained earlier in this chapter, with just 300 qubits in a superposition state, a quantum computer can check as many input combinations as there are atoms in the universe.

It would seem that a quantum computer is perfectly suited for tackling inverse problems.

As another example, let us consider the famous *travelling salesman problem.* Consider a salesman who is travelling around America. His schedule requires him to visit a certain number of specified cities, returning to the city he started from. However, the salesman has complete freedom as to the order in which he visits those cities. The salesman wants to save as much fuel as possible, so he selects to visit those cities in a certain order in an attempt to minimise the total distance (obviously, zig-zagging across America would be a bad idea as it would maximise the total distance).

The example in the following diagram shows the travelling salesman having to visit five cities. The cities have been given distinguishing letters, and the distance (in miles) for all the possible routes between the cities are displayed. Imagine that the salesman starts his journey in city **A**. Can you find the shortest route in which the salesman visits all five cities, before returning to his starting city? I will give the answer later.

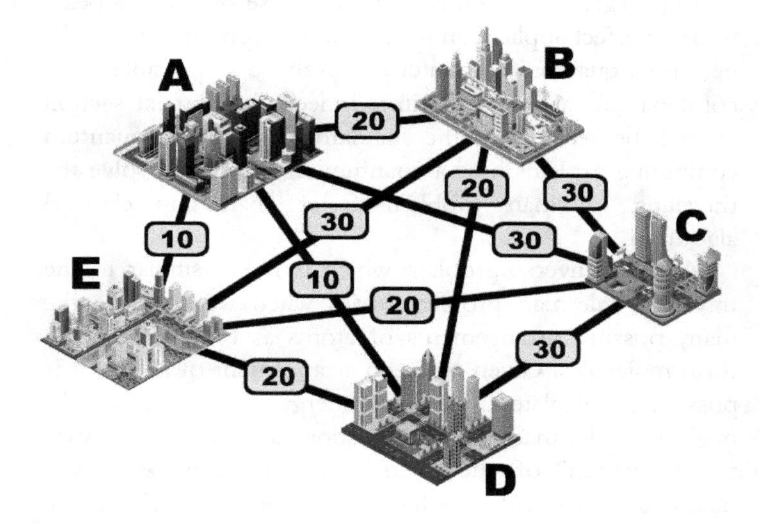

It is clear that in this case we have a forward transformation: given a particular route as input, it is easy to calculate the total distance of the route. But the inverse transformation is extremely difficult: given a particular total length of a route, how do you calculate the route which taken? Or how do you calculate the minimum possible total distance, visiting all cities?

It turns out that there are a huge number of different possibilities. If there are *n* different cities, then the number of different routes is equal to the factorial of (*n*-1). That means it is equal to all the positive integers equal to or less than (*n*-1) multiplied together. As an example, if the salesman has to visit five cities then there are 24 different possible routes (4×3×2×1). That might not sound too bad, but the factorial grows very rapidly as the number of cities increases. If the salesman had to visit just 52 cities, for example, there would be 10^{62} different routes (the number 1 followed by 62 zeroes)! In that case, how do you check all the possible routes?

Once again, this example of an inverse problem appears to be a perfect application for a quantum computer to tackle, because a quantum computer can evaluate all possible input combinations at the same time. Indeed, in the next section we will be seeing that the standard textbook on quantum computing explains how a quantum algorithm can solve the travelling salesman problem faster than any classical algorithm.

Another inverse problem which is rather similar to the travelling salesman problem arises when we consider the many possible arrangements of atoms as they combine to form molecules. Given a certain arrangement of atoms, it is possible to calculate the resultant energy of the molecule. We might consider that energy calculation as being the "forward transformation" of our inverse problem. A molecule will then tend to adopt the configuration of atoms which has the lowest energy: this is known as the ground state (we might think of it as the equivalent to the shortest route for the travelling salesman). But, given the ground state energy, is it then possible to work backwards to determine the arrangement of atoms? This then becomes an example of an *optimization problem* in which we have to try to find the best solution from a range of possible solutions (the travelling salesman problem is another example of an optimization problem in which the total distance has to be minimised). By determining the structure of molecules in this way, new life-saving drugs can be developed, and this is an area in which quantum computing is already showing great potential.

Shor's algorithm

But there is one particular inverse problem which has generated a great deal of attention and has made quantum computing a front page story. It is the inverse problem which involves finding the factors of a number.

It is a relatively simple process to multiply two integers (whole numbers) together, even very large numbers, to produce a larger number. We might think of the initial two numbers as representing the input to our process, and the act of multiplying the two numbers together as being the "forward transformation". It is clearly an easy task for a computer to multiply two large numbers together. However, the inverse problem then arises when we consider the resultant large number and want to work backward to determine the initial two numbers which were multiplied together. The initial two numbers are called the *factors* of the larger number, and the process of determining the two factors is called *factorization.*

It turns out that there is no quick and easy method to factorize a large number. In fact, it is incredibly hard and time-consuming – even for the most powerful classical computer. So, while the forward transformation is easy (multiplying two numbers together), we do not have access to a practical inverse transformation. And that should immediately make you think: "Aha! This sounds like an inverse problem which could be successfully tackled by a quantum computer!"

Indeed, in 1994, the mathematician Peter Shor developed a quantum algorithm for factoring large numbers. Shor's algorithm, utilising the power of quantum parallelism, was far faster than any existing classical algorithm.

For this reason, Shor's algorithm generated a lot of attention and newspaper column space because it threatened to crack the encryption technique – known as RSA encryption – used to secure internet communications. Every time you type "https:" into your internet browser address bar, you are using RSA encryption to encode your credit card details (the "s" in "https" stands for "secure"). If Shor's algorithm could crack RSA encryption, the consequences would be huge.

RSA encryption was first explained in a 1978 scientific paper which introduced us to two fictional characters named Alice and Bob. The names "Alice" and "Bob" (instead of just using A and B, as was usual up to that point) have since become the accepted names which are used in many thought experiments in physics.

To understand RSA encryption, imagine that Alice and Bob live a considerable distance apart. Now imagine Bob wants to send an encrypted message to Alice which only she can read. The conventional method of sending a coded message would require Alice and Bob to share a *key*, which is known only to themselves. Bob would use the key to encode his message, and Alice would then use the key to decode Bob's message. However, transmitting the key between Bob and Alice in advance of the message is inconvenient and a security risk.

Rather amazingly, RSA encryption does not require the secret sharing of a key in advance. To understand how RSA encryption works, first Alice picks two large prime numbers, say, p and q. These two numbers then represent Alice's *private key* which she keeps to herself. Alice then multiplies those two numbers together to form her *public key*. Alice must then publish her public key in a publicly-available directory – anyone can read it. However, Alice's private key still remains completely secure because of the difficulty of factoring large integers: no eavesdropping third party can work out what p and q are – Alice's private key – even if they have access to her public key. Even though the public key contains both p and q multiplied together!

When Bob wants to send a message to Alice, he encodes it using her public key, and then sends the message to Alice. This is entirely secure even if the message is intercepted because the resultant encoded message can only be decoded using Alice's private key. When Alice receives the message from Bob, she decodes it using her private key and reads the message. So encrypted messages using RSA encryption can be sent between Alice and Bob without the requirement for a secret key to be exchanged in advance.

It is clear that if Shor's algorithm could factor large numbers, it could reveal Alice's private key, and RSA encryption – the backbone of internet security – would no longer be secure.

Peter Shor's algorithm is clearly a tremendous achievement. However, it is specific to the problem of factoring numbers. The aim of this book, however, is to present general techniques which IT professionals might use for their own applications, and thereby take quantum computing into the mainstream. Though Peter Shor's algorithm is impressive, I suspect most IT professionals will not be interested in cracking internet security.

With that aim in mind, in the next chapter we will be examining a general technique for solving inverse problems which might be used in a wide range of applications.

The challenge of writing a quantum algorithm

So we have now gained have an understanding of how a quantum algorithm works. Initially, we have a set of qubits which are initialised to be in either state $|0\rangle$ or state $|1\rangle$. Those qubits are placed in a superposition state by applying a Hadamard gate to each qubit. Because they are now in a superposition state, each qubit then becomes both "zero and one at the same time". The qubits then pass through a series of quantum gates, and the values of the output qubits are generated. When those output qubits are measured, we obtain the result of our calculation.

However, the principle of massive parallelism which gives a quantum computer its power also makes a quantum algorithm very difficult to write. Because the qubits are in a superposition state, all possible values of the input data are presented to the quantum computer at the same time. But when we measure the output of the computer, at that moment there is the usual collapse of the quantum state, taking the quantum computer out of its superposition state. At this point, the quantum state collapses to a single well-defined value (as described in Chapter Four: "Observing a quantum system").

So, at that point, the output qubits collapse to a single value. If we then trace that output value back through the workings of our quantum computer, then that output value would represent just one particular input value. In other words, the input also suddenly becomes just one value. It is as if your entire quantum computer has suddenly collapsed to become a classical computer! To make matters worse, we have no control over which particular input value the qubits will collapse to – the collapse process is, of course, completely random.

This difficult situation is described by Eleanor Rieffel and Wolfgang Polak in their book *Quantum Computing: A Gentle Introduction.* As Rieffel and Polak say, the act of measuring will "project the final state onto **a single input/output pair, and a random one at that."**

The problem is explained in the following image:

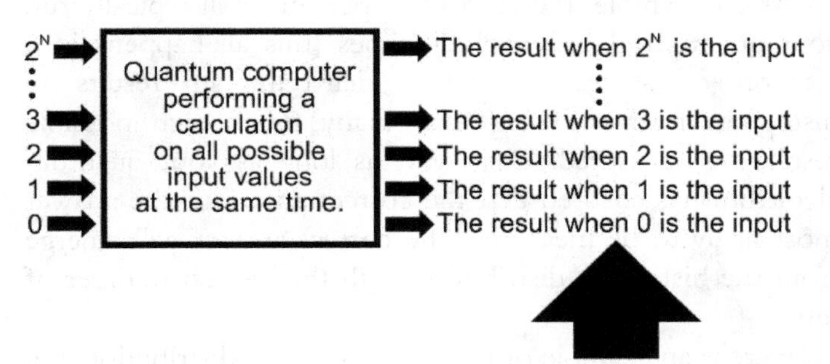

The previous diagram shows the numerous input/output pairings, with only one pairing being chosen at random to represent the output. Once again, this difficult situation is described by Rieffel and Polak: **"Only one result is obtained and, even worse still, we cannot even choose which result we get."**

So how can we ensure that we get the correct value in the output qubits at the end of the calculation? There are so many possible input data values, and we have no control over which input value will eventually be randomly selected when we observe the output. In that case, how can we ensure that the output qubits collapse to the correct value of the calculation?

The situation sounds almost impossible! How on Earth can it be done?

Well, it turns out that it **is** virtually impossible: you can't write a quantum algorithm that gives you the correct answer every time.[13] Instead, **you have to increase the probability of the correct answer being found on the output qubits,** and then rely on the probabilistic nature of quantum mechanics to do the rest. That is the best that you can do.

As an example, the IBM Q Experience will typically run your quantum algorithm 1024 times (this all happens in a fraction of a second) before delivering its results in histogram form showing how many times each possible result was measured. That way, as long as your quantum algorithm has ensured that the correct answer is the answer most likely to be measured, the correct answer will emerge from the histogram distribution with the highest number of hits.

Here is an example of the final histogram distribution in a screen capture from my earlier Hadamard gate video:

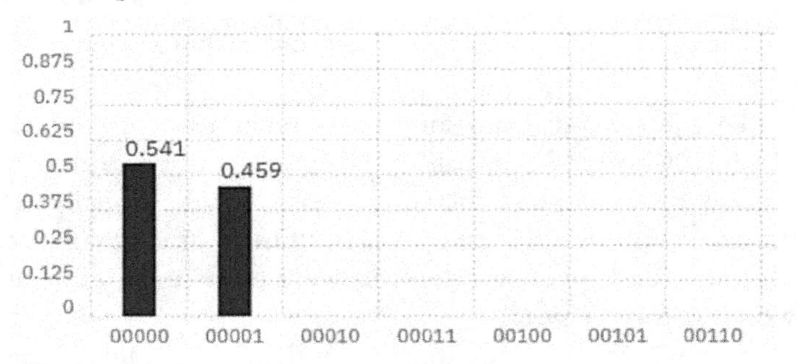

[13] That is not entirely true. There is an algorithm called *Deutsch's algorithm* which gives the correct result every time. Deutsch's algorithm was devised by David Deutsch in 1985 and was the very first quantum algorithm. Unfortunately, it does not do anything of practical use – it was merely devised to show that quantum algorithms can be faster than classical algorithms.

And this is where quantum entanglement plays a vital role. In the discussion of entanglement back in Chapter Seven, it was explained that when we are dealing with entangled qubits, you should no longer think of the qubits as independent units. Instead, you should think of the entangled qubits as representing a single quantum system, **described by a single state vector.** The task of your quantum algorithm is then to manipulate (rotate) that single state vector to align it more closely with the correct answer to ensure that the correct answer is more likely to be observed when the output of your quantum computer is finally measured.

If that sounds rather tricky, well, it is. Fortunately, I am not going to be suggesting that that is what you should attempt to do when you come to writing your own quantum algorithms. I am not going to be suggesting that you start from scratch, and have to devise some ingenious state vector rotation algorithm every time. Instead, in the next chapter – the final chapter of this book – a quantum algorithm will be presented which can be adapted to your own particular needs. In other words, the ingenious method of state vector rotation has been worked-out for you in advance, and you just have to modify one section of the algorithm to match your own particular problem.

(By the way, I promised I would give you the answer to the travelling salesman puzzle presented earlier. If you visit the cities in the order ADBCEA – or the reverse order, of course – then you will travel a total distance of just 90 miles, which is the shortest possible distance.)

11

GROVER'S ALGORITHM

So what we really need is a general quantum algorithm for solving inverse problems. Quantum computing could then be adopted by IT professionals, and the potential of quantum computing could then be finally fulfilled. Fortunately, in 1996 the Indian-American computer scientist Lov Grover devised an algorithm which fits these requirements. The method is now known as *Grover's algorithm.*[14]

Grover's algorithm is often explained by an example using four playing cards. The following image shows four playing cards, one of which is the queen of spades. The cards are then turned face down, and randomly rearranged so that you no longer know which card is the queen. You will see in the diagram that each face-down card has been given an

---

[14] Lov Grover, *A fast quantum mechanical algorithm for database search*,
http://tinyurl.com/lovgrover

identifying number. Your task is to identify which card is the queen.

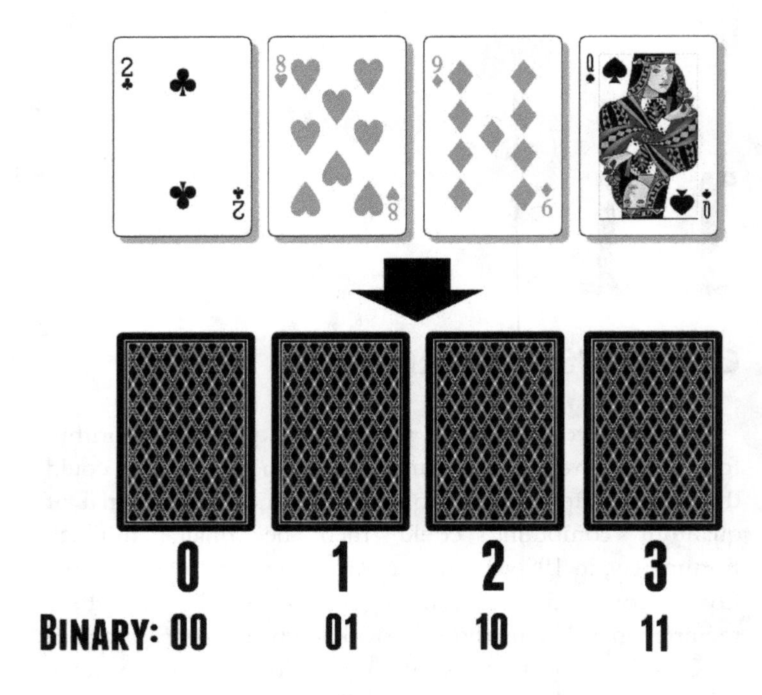

We can express this problem in a similar form to the other problems we have considered so far. We can consider the numbered position of the card (0, 1, 2, or 3) as being the input to a forward transformation. The nature of the forward transformation in this case is very simple: you take that numbered position of the card and you turn that particular card over to see what it is. The output of the forward transformation is then the Boolean value TRUE or FALSE depending on whether or not that particular card turned out to be the queen of spades.

So we have a simple forward transformation (from the question to the answer), but do we have an inverse transformation (from the answer to the question)? No we do not. If you are presented with the answer TRUE, meaning the queen of spades has been found, then from that answer can you determine which card at which numbered position was turned over? No, there is no way to do that (unless the cards were marked).

So, once again, we seem to have a problem with a simple forward transformation but no inverse transformation. And that should make you think, once again, that this is a representation of a problem which could be cracked by a quantum computer.

If the positions and values of the cards were represented by variables in a computer program, then the problem could be translated into a problem running on a computer. We could then use the forward transformation to turn every card over in sequence until we find the queen. We know that process by another name: a *search*. So Grover's algorithm – which can be used to solve this problem of finding the queen – is an example of a *search algorithm*.

Let us now consider how long it would take us to find the queen by a manual search (or, equivalently, a search on a classical computer in which only one card at a time can be turned over).

The position of the queen is completely random, so, initially, the queen can be in any of the four numbered positions. Let us imagine our manual search always starts on the left-hand side. Then, if the queen is located on the extreme left position, we will find it on our first attempt, turning over just one card. Otherwise, if the queen is located in the second-from-left position, we will find it on our second attempt, turning over two cards. Otherwise, if the queen is located third-from-left we will find it on our third attempt, turning over three cards. And, in the worst case scenario, with the queen positioned at the extreme right, we will have to turn over all four cards before we find the queen.

With there being an equal chance that the queen was in any one of the four positions, the average number of cards we will have to turn over before we find the queen is:

$$\frac{1+2+3+4}{4}=2.5$$

So, using a manual exhaustive search (or a search on a classical computer), we will have to turn over an average of 2.5 cards before we find the queen.

But Grover's search algorithm can find the queen by turning over just one card.

Grover's algorithm can speed up an exhaustive search by a factor proportional to the square root of the number of classical operations. In this example, with 2.5 classical operations having to be performed to find the queen, Grover's algorithm promises to find the queen using only the square root of 2.5 number of operations – which is a number close to 1.5. And this is what we find. In fact, Grover's algorithm in this particular case can always find the queen in only one operation, one attempt.

And Grover's algorithm can be applied to a huge range of different search operations. Michael Nielsen and Isaac Chuang have written the standard textbook on quantum computing entitled *Quantum Computation and Quantum Information*. They consider the travelling salesman problem in their book, and present Grover's algorithm as a potential solution:

> *Suppose you are given a map containing many cities, and wish to determine the shortest route passing through all cities on the map. A simple algorithm to find this route is to search all possible routes through the cities, keeping a running record of which route has the shortest length. On a classical computer, if there are N possible routes, it obviously takes a number of*

> *operations proportional to N to determine the shortest route using this method. Remarkably, there is a quantum search algorithm, sometimes known as Grover's algorithm, which enables this search method to be sped up substantially, requiring a number of operations only proportional to the square root of N. Moreover, the quantum search algorithm is* ***general*** *in the sense that it can be applied far beyond the route-finding example just described to speed up many (though not all) classical algorithms that use search.*

Nielsen and Chuang also consider the potential of Grover's algorithm for searching for the factors of numbers, and thereby breaking codes. In order to find the factors of a number, an exhaustive search could progressively examine all the positive integers (whole numbers) smaller than that number, and see if they could be used to divide the larger number – leaving no remainder. If so, the smaller number is a factor of the larger number.

For factoring applications, Grover's algorithm is not as fast as Shor's algorithm, and so Grover's algorithm could not be used to break RSA encryption in a reasonable time.[15] However, it is known that Grover's algorithm could be used to break the widely-used AES encryption scheme which does not use integer factorisation. Grover's algorithm could break

---

[15] Shor's algorithm is **exponentially** faster than the best known classical algorithm, for example 2^n, where n relates to the size of the number being factored, whereas Grover's algorithm only offers a **quadratic** speedup, for example, n^2. The squared term arises because, as stated in the text, Grover's algorithm only takes the square root of the time of a classical algorithm.

the code in the square root of the time it would take a classical computer.

Solving a problem in the square root of the time might not sound impressive, but it can make a tremendous difference. An article on the Betanews technology website revealed the true implications of the "square root" speed-up achieved by Grover's algorithm:[16]

> *If a classical computer needs to search 2^{56} possible keys to be guaranteed to crack DES encryption, a quantum computer running Grover's algorithm only needs to do 2^{28} searches. This is easier to understand when written in conventional notation:*
>
> *Classical computer: 2^{56} searches = 72,057,594,037,927,936*
>
> *Quantum computer: 2^{28} searches = 268,435,456*
>
> *When measured in terms of time, assuming both computers can search at the same speed: if it takes a classical computer one day to crack a particular 56-bit encryption, it would take the quantum computer just 0.322 milliseconds – or one thousandth the blink of an eye. And if it took a classical computer one year to crack 64-bit encryption, it would take a quantum computer 7.3 milliseconds.*

There is another benefit of Grover's algorithm. As described by Eleanor Rieffel and Wolfgang Polak in their book *Quantum Computing: A Gentle Introduction*: "Grover's

---

[16] Linus Chang, *How secure is today's encryption against quantum computers?*,
http://tinyurl.com/squarerootspeed

algorithm is simpler and easier to grasp than Shor's, and has an elegant geometric interpretation."

It is clear, then, that Grover's algorithm represents a method of solving a wide variety of inverse problems. This is what we have been looking for.

So let us now consider the structure of Grover's algorithm in full detail (although, you might be relieved to know that the bulk of the complicated detail has been moved to the two appendices at the back of this book). I suspect for many IT professionals and researchers looking to solve their own problems on a quantum computer, this will be the material they need to know.

As a reward, later in this chapter we will program an example of Grover's algorithm on the real IBM Q Experience quantum computer.

Yes, we'll be programming a real quantum algorithm on a real quantum computer!

The Oracle

The structure of Grover's algorithm can be split into two stages. Here is a block diagram of Grover's algorithm:

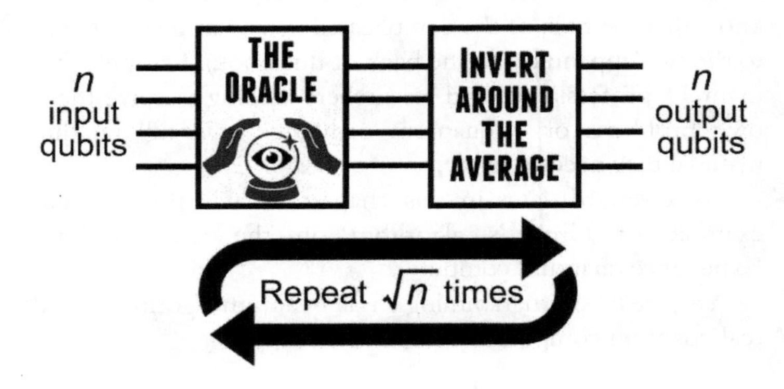

You can see from the previous diagram that Grover's algorithm is composed of two main blocks called "The Oracle" and "Invert around the average". I am going to jump slightly ahead at this point and show you how we will be coding these two blocks in the IBM Q Experience later in this chapter. The two blocks are going to look like this:

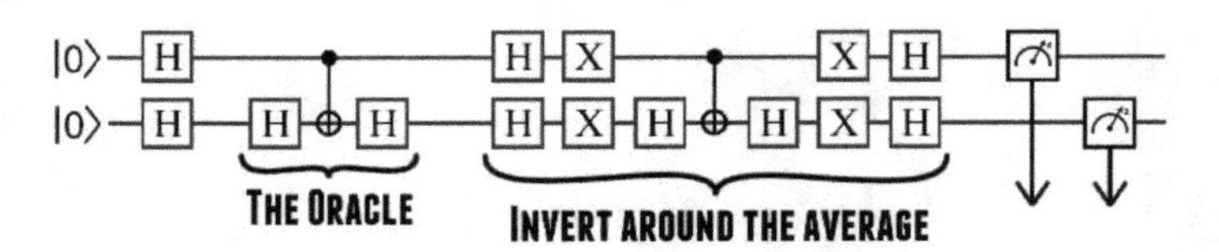

We are going to be using Grover's algorithm to find the position of the queen of spades in the previous playing cards example. As described earlier, this is a search operation – we have to search to find the queen.

There are four possible positions for the queen, with the positions numbered 0, 1, 2, or 3. The binary equivalent of those decimal numbers would be 00, 01, 10, and 11. So the binary numbers have two bits. Hence, our quantum algorithm is going to have two input qubits which will be capable of representing all possible positions of the queen (the previous block diagram of Grover's algorithm can be seen to be based on two qubits).

In this section we will consider the first part of Grover's algorithm, which, as you can see, is called the Oracle.

The first part of Grover's algorithm is a self-contained "black box" of code, a code block which – in computer science terminology – is often referred-to as an "oracle". The term "oracle" seems rather strange, but this is chosen because the unit is defined only in terms of what it does: its inputs, and its outputs. We are not particularly interested in the internal workings, how it derives those outputs from its inputs (this is why it is often called a "black box" – we do not care what happens inside it). Or, to be more specific, its internal workings are not defined, and it is left to the programmers to design how it works internally.

So this "black box" is rather similar to a human "oracle", a visionary or seer with a crystal ball, who you ask a question and who miraculously gives you the correct answer (I seem to remember one of the characters in the *Matrix* movie was called The Oracle – a lady who provided the answers for Keanu Reeves). You don't know how the oracle has done it, you don't know the internal workings – you are just glad it works.

So the first part of Grover's algorithm is an oracle. Let us now consider what the oracle has to do.

The oracle is presented with the n input qubits in a superposition state, which represents all possible input combinations (for two qubits, this would be the four possible states 00, 01, 10, and 11). The oracle then has to perform the following two tasks:

Oracle Task Number One

The first task for the oracle is that it has to consider the input value (00 or 01, etc.) and perform a calculation to see if that value satisfies the particular search problem. In this playing card example, the oracle has to decide if the number represents the correct position of the queen of spades. **So the key task of the oracle is that it has to recognise when a specific combination of input qubit values solves a problem.**

If you wanted to adapt Grover's algorithm to solve your own problems, you would have to adapt this oracle for your own problem. You would have to perform some calculation on the input value in order to test to see if the input value satisfies your search problem. Is it the number you are looking for? For example, is the number a factor of a larger number? Does it divide perfectly into the larger number? If so, you've cracked the code! And, of course, because you are being presented with your input qubits in a superposition state, your calculation will be working on all possible combinations at once. Like magic!

Oracle Task Number Two

The second task for the oracle is that when it finds the solution it is looking for, it has to "mark" the solution. According to Grover's algorithm, this marking is performed by inverting the phase of the state by rotating it by 180°. If you rotate anything by 180°, you are making it point in the opposite direction. So another way of interpreting this "marking" operation is that it puts a minus sign in front of the state vector.

Let us now consider how we can create a quantum oracle on the IBM Q Experience.

In our example of finding the queen of spades, we note that in our earlier diagram we placed the queen at the extreme right-hand position of the four cards. So we want to create our oracle to detect when this numbered position is input. You will notice on the earlier diagram of the playing cards that the right-hand position was given the number 3, which is equivalent to the binary number 11. So we want our quantum oracle to mark the input state when both input qubits are in state $|1\rangle$ – and not when the qubits are in any other state. In other words, we need to invert the phase only when both input qubits are in state $|1\rangle$.

It turns out that this phase inversion can be performed by a very simple circuit. It can, in fact, be done with a single CNOT gate and a couple of Hadamard gates. The following diagram is the actual circuit diagram for the Grover's algorithm oracle:

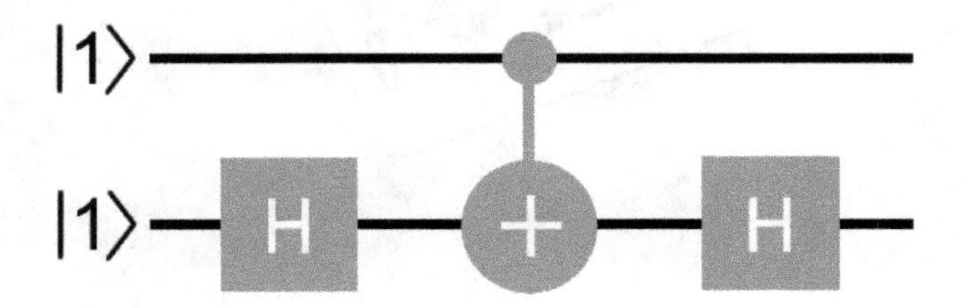

In the example shown in the previous diagram, both inputs are set to state $|1\rangle$. As just explained, this is the condition which we want to result in the phase inversion.

The details of how this oracle circuit was designed are fairly complicated, so they has been moved to Appendix One at the back of this book.

But we now have the design for our quantum oracle circuit, so we are getting closer to finding our queen …

Inversion around the average

It is now time to move to the second stage of Grover's algorithm. And, as we shall see, it is quite ingenious.

To recap, the queen of spades has been found and the state representing the position of the queen (11) has been marked. But what do we do next with this marked state? How do we ensure that when we look at the output of our Grover's algorithm, the marked state will appear as the most likely solution with high probability? We need to find some way of amplifying the marked state.

Remember, we marked the state by inverting the phase of that state so it points in the opposite direction to the phases of all the other states. This situation is shown in the following diagram showing the phases of all the states. You can see that all the phases point up except for the single phase of the marked state which was inverted and therefore points down:

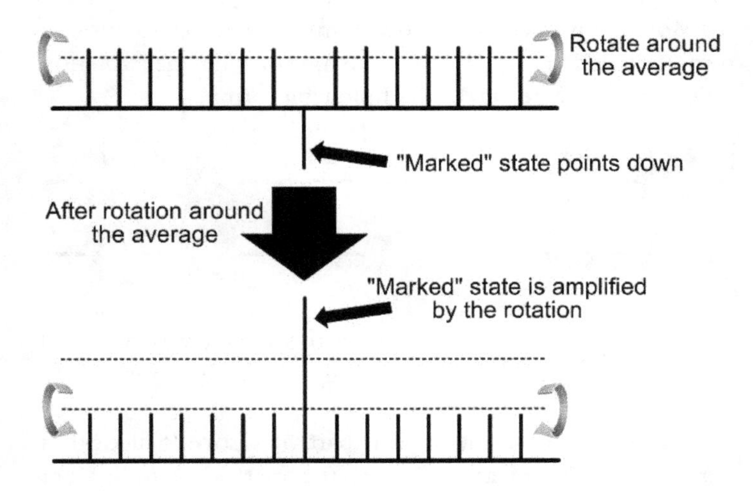

In the top part of the diagram you will also see a thin dashed line. This represents the **average** value of all the phases. Because almost all the phases are pointing upward, this average value is quite large, almost precisely aligned with the tips of all the lines which are pointed upward. The only thing which is dragging the average value down very slightly is the value of the single marked state which is pointed downward.

The next step is very clever. We rotate **all** the states by 180° around the average value, the dashed line. You can see the effect in the lower part of the previous diagram. Because the value of the upward-pointing states was very close to the average value, the effect of the rotation does not affect them very much. However, as you can see in the previous diagram, the tip of the downward-pointing state was very far away from the average value – approximately twice as far away as all the other upward-pointing states. So when we rotate all the states around the average value, the downward-pointing state now gets amplified to more than twice its original length, and now points upward – as shown in the lower part of the previous diagram.

Moving on to consider how to implement this rotation, it turns out that the final circuit for performing the "rotation around the average" takes the following form:

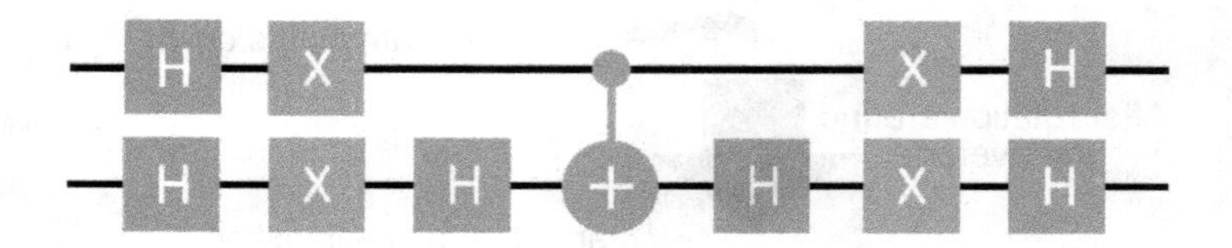

Once again, the details of how this circuit was designed are quite complicated and so they have been moved to Appendix Two at the back of this book.

And that is how the second part of Grover's algorithm works. The amplification boosts the marked state so that when the output of the quantum computer is observed, it is more likely to collapse to the correct state, revealing the position of the queen of spades. Usually several iterations have to be performed (the square root of n times) until the correct state receives sufficient amplification to ensure that it will be selected with high probability.

Programming a real quantum algorithm on a real quantum computer

So we have discovered the complete structure needed to implement Grover's algorithm. Now let us program this using the IBM Q Experience.

Here is the final complete block diagram:

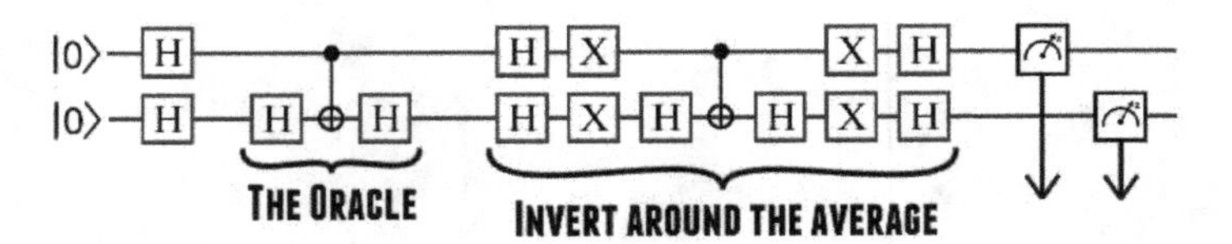

This circuit diagram for implementing Grover's algorithm can be programmed directly into the IBM Q Experience. Give it a go and see if it works!

I am actually ahead of you because I have already tried it, and I have created a video of my efforts. Here is a link to the video:

http://tinyurl.com/magicquantum

And here is a still from the video:

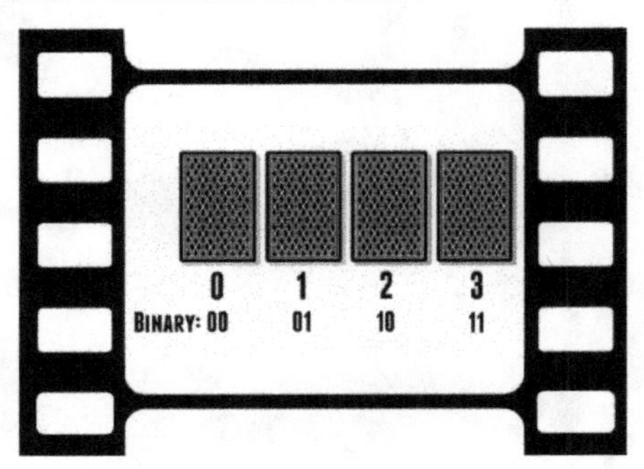

The revealing of the queen of spades at the end of the video represents perhaps the most technologically-advanced magic trick in history.

Some final thoughts

I hope you have enjoyed the book.

When I wrote this book, I had three goals in mind.

Firstly, I was aware that there was a general lack of clear, easy-to-understand technical information about quantum computing. Therefore, I wanted to write an accessible book which would capture the essential concepts of quantum computing, hopefully simplifying a complicated subject.

Secondly, I wanted the book to appeal to software engineers and scientific researchers – anyone who might find quantum computing useful in their line of work. I wanted to explain the methods which they would most likely find useful to solve their particular problems. It does appear that Grover's algorithm is currently the most flexible method which might be applicable to a wide range of inverse problems.

Finally, I hoped to make the book entertaining and informative for people who have no intention of ever going near a quantum computer. If I have succeeded in that goal alone, then I will consider this book to have been a success.

APPENDIX ONE: THE ORACLE

This first appendix explains how the oracle circuit for the Grover's algorithm was designed. This is fairly complicated, so please only read it if you are interested in the gory details, and are maybe interested in extending it for your own applications.

Otherwise, please skip it.

Let us remind ourselves of the oracle circuit diagram which was presented in the previous chapter:

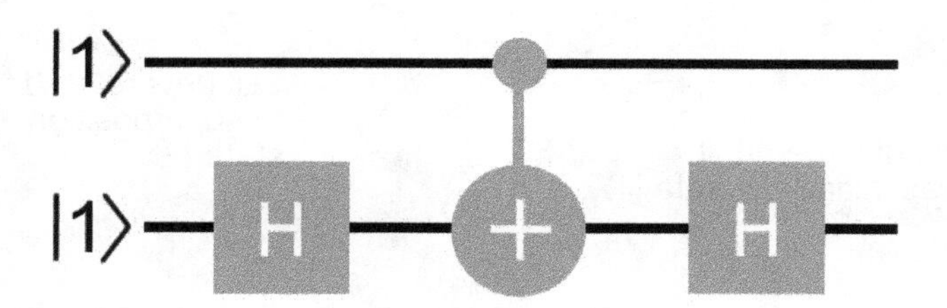

Like all quantum circuit diagrams, we read the diagram from left to right.

You can see that the target qubit – the lower qubit – passes through the first Hadamard gate. The qubit state vector then gets rotated according to the usual behaviour of the Hadamard gate (see Chapter Eight for a reminder). So, as you can see in the following diagram, the qubit state vector gets rotated around the diagonal dashed line and thereby enters a superposition state:

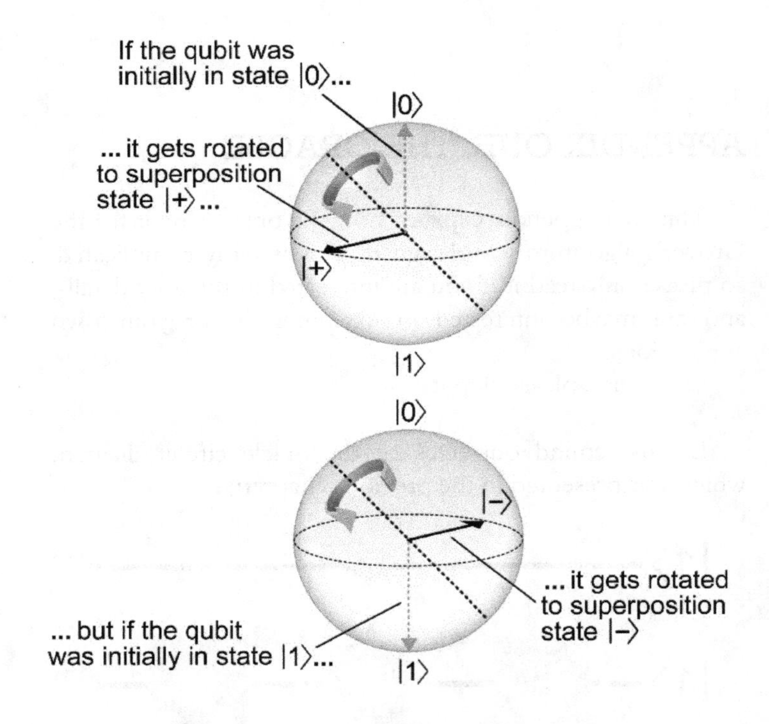

But you can see from the previous diagram that the direction of the state vector in the superposition state is different depending on whether the initial state was $|0\rangle$ or $|1\rangle$. If the initial state was $|0\rangle$ (state vector pointing up), then the state vector gets rotated to a position pointing toward the left and to the front (see the previous diagram). This is called superposition state $|+\rangle$ (the "plus state"). However, if the initial state was $|1\rangle$ (state vector pointing down), then the state vector gets rotated to a position pointing toward the right and to the back. This is called superposition state $|-\rangle$ (the "minus state").

In this playing cards example, we are passing a target qubit in state $|1\rangle$ to the Hadamard gate. So, in this case, the qubit enters the superposition state $|-\rangle$.

Expressed mathematically, the superposition state $|+\rangle$ is the usual equal mix of $|0\rangle$ and $|1\rangle$. However, the superposition state $|-\rangle$ is an equal mix of $|0\rangle$ and $-|1\rangle$ (the $|1\rangle$ now has a negative sign in front it). That explains why this superposition state vector is pointing in the opposite direction to the usual direction.

For the next stage of the oracle, you can see that the CNOT gate is then applied to this qubit. Because the control qubit of the CNOT gate is also in state $|1\rangle$, the CNOT gate performs its usual "flipping" operation on the target qubit. This has the effect of flipping the values of the $|0\rangle$ and $|1\rangle$ terms of the state vector, changing $|0\rangle$ to $|1\rangle$, and vice versa. For a state vector in the superposition state $|+\rangle$, both of those terms are positive so flipping the terms has no effect. But for the superposition state $|-\rangle$ we have in this example, as mentioned earlier, the $|0\rangle$ term is positive but the $|1\rangle$ term has a negative sign in front of it. Therefore, flipping the two terms has the effect of placing a negative sign in front of the entire qubit state, as explained in the following diagram:

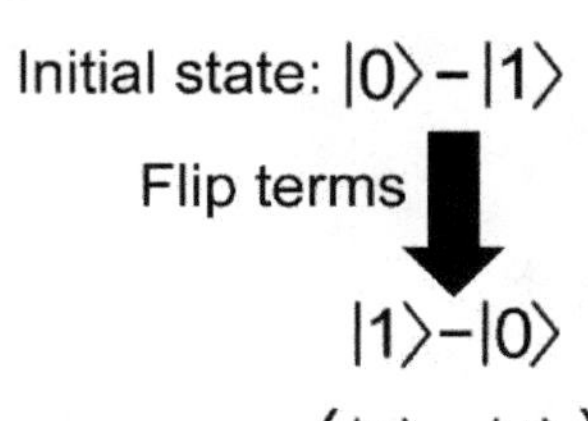

Which is the same as: $-(|0\rangle - |1\rangle)$
(i.e., the negative of the initial state)

And that is just what we have been looking for! If you remember, we wanted to find a way of inverting the state, which, as stated earlier, is the same as putting a negative sign in front of the state. And that is just what we have done with our CNOT gate.

So we have achieved the phase inversion we desired. And this phase inversion is caused when the following two conditions are true:

1) The lower of the two qubits in the circuit diagram must be in state $|1\rangle$ so that the qubit enters the superposition state $|-\rangle$.

2) The higher of the two qubits in the circuit diagram must also be in state $|1\rangle$ so that the CNOT gate does its "bit-flipping" operation.

So both input qubits must be in state $|1\rangle$ for the phase inversion to take place. And, if you remember, that is precisely how we wanted our quantum oracle to behave: we wanted our quantum oracle to perform a phase inversion ("marking" the state) when both of the input qubits are in state $|1\rangle$.

We have therefore found our quantum oracle!

APPENDIX TWO: ROTATION AROUND THE AVERAGE

This appendix describes how the "rotation around the average" quantum circuit in Grover's algorithm was designed. Once again, this is tricky stuff, so only read it if you need to.

We know how to do rotations, as we have just used the method in the quantum oracle. We will be reusing the same piece of code, and it will form the centrepiece of our new "rotation around the average". Here, once again, is the circuit diagram which performs the rotation:

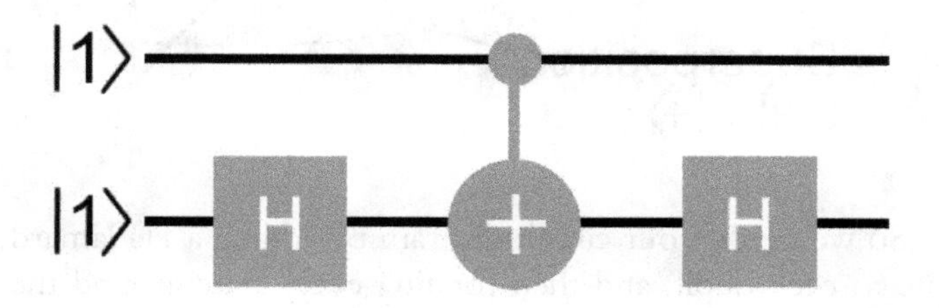

Let us now extend this circuit diagram to perform the "rotation about the average".

You might think it would be a tricky calculation to find the average value of all the states, but it is actually extremely easy. An average might be considered as being halfway between two extreme values. Building on that idea, if we have two qubits, one in state $|0\rangle$ and the other in state $|1\rangle$, the half-way value would be the superposition state, halfway between state $|0\rangle$ and state $|1\rangle$. So the average, in this case, is defined by the superposition state. In fact, it can be shown that – no matter how many states you have – the average is always the equally-weighted superposition state.

So that makes things a lot easier: to perform the "inversion around the average" we need to rotate all the states around the superposition state $|+\rangle$. But there is an easier way of achieving that. Remember, if a Hadamard gate is applied to a superposition state $|+\rangle$, it gets rotated to state $|0\rangle$, as shown in the following diagram:

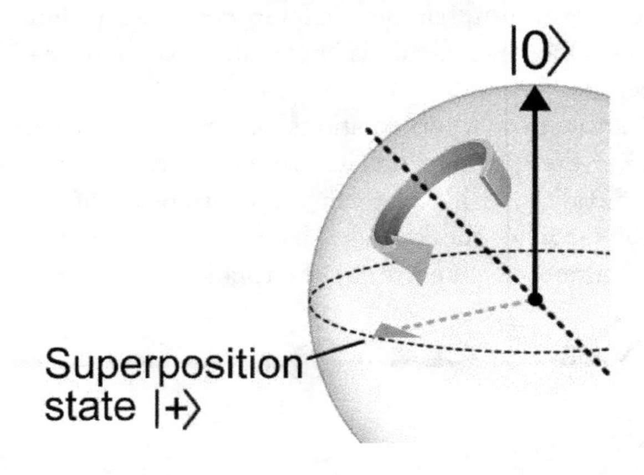

So we modify our circuit diagram by adding a Hadamard gate to each qubit, and then rotating every state around the state $|0\rangle$.

Actually, we don't have to rotate every state – there is an easier method. It is clear that we would get exactly the same result if we kept all the states the same, and just rotated the $|0\rangle$ state in the opposite direction. The final configuration of states would be exactly the same as if we had rotated everything around $|0\rangle$. So that is what we do: we just have to rotate the $|0\rangle$ state.[17]

We also need to add a couple of X gates (quantum NOT gates) because we know from our previous experience with the oracle that the circuit we used only performed the rotation when both qubits were in state $|1\rangle$, but we want the rotation to occur for state $|0\rangle$.

We end up with the following circuit to perform the "rotation around the average":

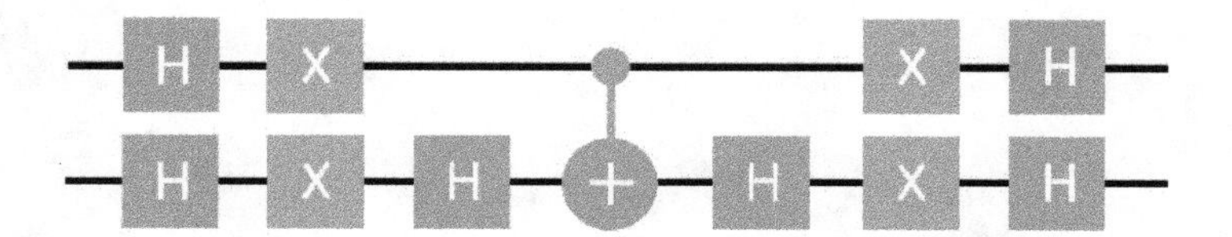

[17] This clever method is described by Dayton Ellwanger in one of his excellent (though technical) instructional videos: **http://tinyurl.com/daytonvideo**

PICTURE CREDITS

All photographs are public domain unless otherwise stated.

Photograph of IBM 601 Multiplier is by Sandstein and is provided by Wikimedia Commons.

Photograph of Richard Feynman is by Tamiko Thiel and is provided by Wikimedia Commons.

Photographs of the IBM quantum computer is courtesy of IBM Research.

HIDDEN
INPLAINSIGHT11
The logic of consciousness
AVAILABLE NOW

CPSIA information can be obtained
at www.ICGtesting.com
Printed in the USA
LVHW052143260623
750878LV00032B/701

9 781726 017572